SpringerBriefs in Geography

AF793723

For further volumes:
http://www.springer.com/series/10050

Arie S. Issar

Strike the Rock and There Shall Come Water

Climate Changes, Water Resources and History of the Lands of the Bible

Springer

Arie S. Issar
The J. Blaustein Institute for Desert Research
Sede-Boqer Campus
Ben-Gurion University of the Negev
Beer-Sheva
Israel

ISSN 2211-4165 ISSN 2211-4173 (electronic)
ISBN 978-3-319-01936-9 ISBN 978-3-319-01937-6 (eBook)
DOI 10.1007/978-3-319-01937-6
Springer Cham Heidelberg New York Dordrecht London

Library of Congress Control Number: 2013947784

© The Author(s) 2014
The facts and opinions expressed in this work are those of the author and not necessarily those of the publisher.
This work is subject to copyright. All rights are reserved by the Publisher, whether the whole or part of the material is concerned, specifically the rights of translation, reprinting, reuse of illustrations, recitation, broadcasting, reproduction on microfilms or in any other physical way, and transmission or information storage and retrieval, electronic adaptation, computer software, or by similar or dissimilar methodology now known or hereafter developed. Exempted from this legal reservation are brief excerpts in connection with reviews or scholarly analysis or material supplied specifically for the purpose of being entered and executed on a computer system, for exclusive use by the purchaser of the work. Duplication of this publication or parts thereof is permitted only under the provisions of the Copyright Law of the Publisher's location, in its current version, and permission for use must always be obtained from Springer. Permissions for use may be obtained through RightsLink at the Copyright Clearance Center. Violations are liable to prosecution under the respective Copyright Law.
The use of general descriptive names, registered names, trademarks, service marks, etc. in this publication does not imply, even in the absence of a specific statement, that such names are exempt from the relevant protective laws and regulations and therefore free for general use.
While the advice and information in this book are believed to be true and accurate at the date of publication, neither the authors nor the editors nor the publisher can accept any legal responsibility for any errors or omissions that may be made. The publisher makes no warranty, express or implied, with respect to the material contained herein.

Printed on acid-free paper

Springer is part of Springer Science+Business Media (www.springer.com)

Preface

The book presented hereby is an updated and revised version of my book *Water Shall Flow from the Rock—Hydrogeology and Climate in the Lands of the Bible*, published by Springer-Verlag in 1990. The reason for updating and revising and not just reprinting of the book is due, in the first place, to the enormous quantity and quality of the new data added since 1990. Another reason is the better understanding achieved since the book was first published, of the role which climate changes played in shaping the history of the Lands of the Bible. Albeit the fact that the impact of climate change was touched upon rather widely in the first book, still, without interweaving the new data and its interpretation, the picture would have remained partial.

Other books which I published report on these changes and their impact. In 2003, *Climate Changes during the Holocene and their Impact on Hydrological Systems* was published by Cambridge University Press. Later with Dr. M. Zohar, *Climate Change—Environment and Civilization in the Middle East* was published by Springer in 2004 and in 2007 with the same co-author *Climate Change—Environment and History of the Near East* was also published by Springer. The papers as well as chapters in books focusing on a certain period or an area or both will be cited in the following chapters.

This book takes a stand in the dispute between scholars, historians, and archaeologists, investigating the history of the people of Israel, with regard to the factual background of the stories of the Pentateuch. This dispute reached headlines after the publication of Finkelstein and Silberman's book *The Bible Unearthed*, wherein the authors proclaim that: "[A] careful critical perusal of this work—which certainly has much to say about both archaeology and the biblical writings—reveals that we are dealing very largely with a work of imaginative fiction, not a serious or reliable account of the subject" (p. 464). The author of this book disagrees with Finkelstein's and Silberman's statement. Although he did not dig archaeological sites in the Sinai and Negev, as these archaeologists did, he lived for quite a few years, in these deserts, investigating its surface and subsurface water resources. Many times he camped with the local Bedouin, learning from them about their way of life, especially how to locate a meager supply of water, sometimes brackish and sometimes bitter, in the desert. He learnt from them where and how to hit the rock for the water to come out, or where to dig a well to obtain fresh water. Many times he ate with them their unleavened *pita*, baked on embers, and grilled quail. Many times he could fetch from his knapsack his copy of the Bible and read a chapter describing the episode which he experienced.

These experiences brought the author to the conclusion that many stories of the Bible contain an historical core, which a scientist knowing and understanding the natural environments of the region, can bring to light and share with an interested audience.

Thanks are due to Dr. Matanya Zohar, a partner in writing the books mentioned above which were published by Springer Publishing House. Zohar's extensive knowledge in the field of the archaeology of the Land of Israel and his comments were very helpful. Thanks to Yaakov Issar for his help in translating from Hebrew to English and last but not least to Margalit, wife and partner for her help and philological criticism. Thanks are due to Mrs. Marcia Ruth for editing the book.

Arie S. Issar

Contents

Chapter 1
Caves, Dripstones and Soils Tell About Past Climates

Abstract This book is written for the general audience, and its focus is on groundwater, namely water which flows underground in the openings of the layers, in either pores, fractures or caves. This water becomes surface-water when it finds an outlet, either naturally as springs, or artificially as wells. The study of the history of water-works in the arid and semi-arid parts of the globe led the author to conclude that it provides understanding of the impact of climate changes on the history of the inhabitants of these regions.

Most people encounter groundwater when they visit caves which were formed due to the solution of carbonate rocks, either limestones or dolomites. The solution processes are due to the water becoming slightly acid, after it dissolves carbon dioxide gas which is part of earth's atmosphere. This water penetrated into the carbonate rocks through cracks and fissures which are produced by the breaking of the rocks either due to faulting or folding. Over time the crevices expanded in all direction forming caves from whose ceilings water penetrating through the crevices and solution channels continues to flow, either as drops or spurts. Losing the dissolved gas in the cave the water deposits the dissolved carbonate as dripstones in the shape of stalactites descending from the ceiling and stalagmites growing from the floor of the cave (Fig. 1.1).

Caves in which these dripstones are found, especially caves in arid areas, teach us about periods in which the area was more humid. Yet, not only the space of the caves and the existence of the stalactites and the stalagmites contain a lot of information about the changes of the climate, but as will be demonstrated in the following paragraphs, these calcareous pillars are a treasure of information about climates of the past, waiting to be deciphered.

One such case of deciphering began in 1968 and will be discussed in detail in the following chapters. At that time the author was an employee of the Geological Institute of the Government of Israel in Jerusalem. An engineer in charge on the blasting of a new quarry west of Jerusalem telephoned one of his colleagues in the institute, and notified him that a blast in the wall of the quarry had opened an entrance to a cave. The engineer asked for a geologist to come to investigate the extent of the cave.

A. S. Issar, *Strike the Rock and There Shall Come Water*, SpringerBriefs in Geography,
DOI: 10.1007/978-3-319-01937-6_1, © The Author(s) 2014

Fig. 1.1 Stalagmites and stalactites at the Soreq Cave, west of Jerusalem. *Inset* a cross-section of the stalagmite

The colleague recruited a few geologists, including the author, to help him carry out the job. Equipped with ropes and flashlights, the group drove to the quarry and crept through the narrow opening into the dark cave. In the light of the flashlights appeared before their eyes stalactite pillars, curtains of dripstones and of stalagmites that amazed the geologists with their beauty. They noticed that some of the dripstones had collapsed as a result of the explosions. They left the cave pleading with the engineer to stop the explosions in the area of the cave, and immediately requested to all concerned that the cave will be declared a Conservation Site. The goal, indeed, was achieved, and the Absalom Cave, or the Soreq Cave, as it is known in the scientific literature, became a tourist attraction and for us a treasure of information about past climates.

Here, obeying the precept of "honor to whom honor is due", for opening the heavy cover of the container of the treasure, we are obliged to Professor Miriam Bar-Matthews and her colleagues from The Geological Survey, The Hebrew University and The Weizmann Institute for the environmental isotope analysis of the rings of the stalagmites of the Soreq Cave and the paleo-climates interpretation.

The principle behind stalagmites being a source of data is that these pillars of stone are formed ring after ring. This happens during each year in which there was enough rain to saturate the soil cover, if there is any, and penetrate the crevices to the underground, not being sucked back upwards totally by the roots of any vegetation and evapo-transpirated by its leaves. Once the water penetrated below the reach of the roots and dripped into the cave, part of this water will evaporate and the calcium carbonate which was dissolved in it will be deposited in the shape of a ring of the dripstone. The researchers investigated the chemical and isotopic composition (mainly Oxygen 16/18) of every few rings. The age of the rings was determined according to their isotopic ratios of uranium–thorium. These isotopes are the most reliable and accurate age determinants for carbonate sediments like speleothems. The age of the sample is based on the difference between the initial ratio of $^{230}Th/^{234}U$ and the one in the sample being dated. The method assumes that the sample does not exchange ^{230}Th or ^{234}U with the environment

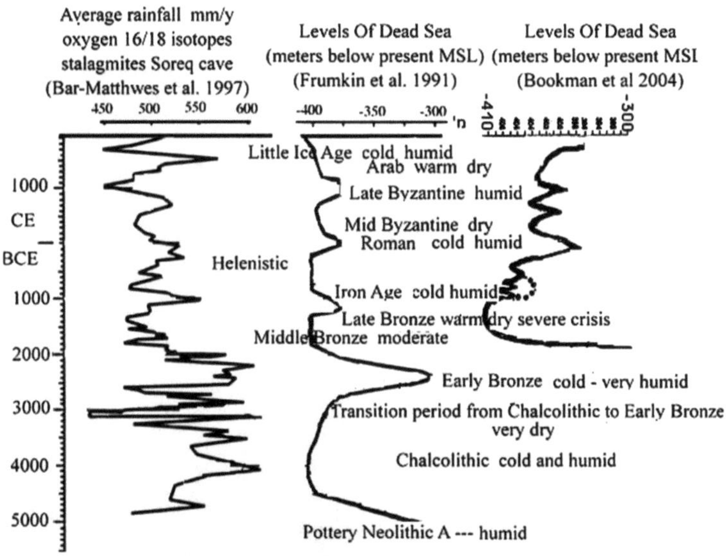

Fig. 1.2 Correlation of data of the isotopic composition of the stalagmites from the Soreq cave with the Dead Sea levels during the last 7000 years

While investigating the isotopic composition of the rings, the relation between the composition of the water dripping into the cave and that of the local rainfall on the surface above the cave were also analyzed and the changes were taken into consideration (Fig. 1.2).

The results of these analyses are presented by the curve on the left side of Fig. 1.2 (Bar-Matthews and Ayalon 2011). The second curve from the left presents the levels of Dead Sea during the last 6000 years. The levels were mapped in the caves of Mount Sodom by Professor Amos Frumkin and his partners (1991, Bar-Matthews et al. 1997, Bar-Matthews, Ayalon 2011). Mount Sodom is a salt plug located at the southwestern edge of the Dead Sea (Fig. 1.3).

The caves or 'chimneys' (Fig. 1.4) were created by the melting of the salt of Mount Sodom by rainwater, which occurs from time to time in this arid region, especially during cold and wet periods.

Frumkin's and his colleagues' investigations found that when the Dead Sea level was higher, its water penetrated into these caves and left behind it the markings of a coastline, along which were deposited pieces of trunks of trees which had been swept to the Dead Sea by floods and floated on the lake's water. The remains of the trees were dated by Carbon 14 and used to date the ancient levels. After the sea retreated the outlet of the caves was opened. The opening enabled the rainwater flowing into the caves from the roof of the salt-plug to erode channels in the floor of the caves, down to the lowest level of the sea. As Mount Sodom is located at the southwestern part of the Dead Sea, it turned out that in some periods the level was lower even than the bottom of the present southern part of the lake. This indicates that in those periods the southern part of the Dead Sea dried up completely.

Fig. 1.3 Mount Sodom at the southwestern edge of the Dead Sea (Courtesy of NASA)

Still, it must be taken into consideration that the information received from the caves is limited to only the altitude of the highest and lowest levels of the lake.

The third curve on the right of Fig. 1.2 is by Dr. Revital Bookman et al. (2004). This investigation was also based on mapping of the ancient high levels of the Dead Sea as exposed as shorelines on the cliffs bordering the lake on the western side.

Fig. 1.4 A 'chimney' in the salt of Mount Sodom (Photo by A. Frumkin)

From the correlation of the three curves it can be concluded that the highest levels of this lake, which also correspond to a period of high precipitation and low evaporation, can be correlated with the information obtained from the curve of the isotopic composition of the stalagmite rings of the Soreq Cave.

Correlating the curves showing the impact of climate changes of the past on the precipitation over Israel to that of the humidity of Anatolia to that of the advance and retreat of the glaciers of Scandinavia as it was mapped and dated by Karlen in 1991, shows that indeed most of the climate changes that Israel has undergone were global. Thus the curve of humidity of Lake Van in Anatolia (Lemcke and Sturm 1997), which belongs, like Israel, to the Mediterranean climate regime, is rather similar to that of the Soreq Cave and Dead Sea, while humid periods are parallel to periods of the advance of glaciers in Scandinavia (Fig. 1.5).

As will be demonstrated in the following chapters, the primary assertion of the present author's former book, *Water Shall Flow from the Rock,* that climate changes did affect the Middle East and had an impact on its history, is maintained. Also retained is the conclusion that cold periods were humid, during which the inhabitants enjoyed "rain in due season" which brought "the land to yield her increase, and the trees of the field to yield their fruit", while warm years brought droughts and famine.

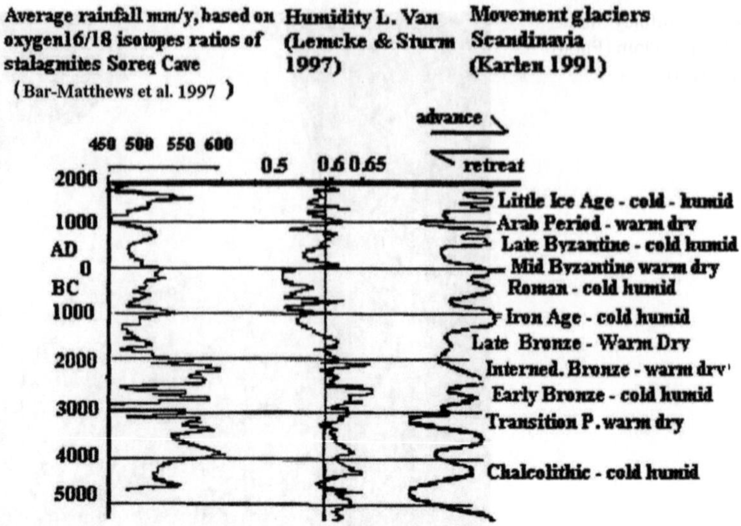

Fig. 1.5 Correlation of climate change curves: Soreq cave, Israel, Lake Van, Anatolia, Glaciers' movement—Scandinavia

The reaction of the inhabitants to the climate changes were according to the severity of these changes and thus their impact on the environment. Once rains became torrential and exceeded the capacity of the soil to absorb them and the capacity of the streams to transfer them to either lake or sea, the land was flooded, causing loss of human and animal life and destruction of habitats. It then took years to reclaim land and property. On the other hand droughts caused famine and forced people to abandon their lands and seek refuge either in the more humid part of their countries or even migrate to another country less affected by the dryness.

Following the order of chapters of the book *Water Shall Flow from the Rock* the present book will present the history of the impact of climate changes on the history according to the order of the events as told by the Bible starting from Genesis to Exodus and the crossing of the Desert of Sinai and Negev.

The reader should, however, bear in mind that the author is not an archaeologist, or a scholar of theology, nor an expert in linguistics or in Bible research. He is a geologist investigating the water resources (hydro-geologist) of arid regions and during the last two decades investigating the impact of climate changes on the natural and human environments. These changes and their impacts on hominid societies started from the beginning of the Quaternary i.e. about a million years before the present (B.P) to present days. At the same time, until he finished high school, the author has been educated by pious parents and teachers to whom the Bible was a holy book to be studied and followed. Thus he became rather well versed in the Bible in its original Hebrew and later in its English translation.

During his career, the author, through his studies of the water resources of arid and semi-arid regions of most parts of the globe, had the opportunity to study the hydrogeology and engineering aspects of ancient water works. Such as the water supply systems of the ancient cities of Canaan, Philistia and Judea, the chain of wells in Iran and the Persian Empire, the ancient water works in Spain and its colonies like Mexico, arid part of Argentina, Chili, etc. The study of the ancient works led the author to study of the translations of Mesopotamian mythologies and documents in which there were interesting references to water works, in some of which he could even follow the linguistic relation between the Babylonian Akkadian version of Enuma Elish and biblical Hebrew, as both are Semitic languages.

A few times, while writing this book, he found himself asking the question whether as a layman in all these sciences of antiquity he does not sin with superficiality. The answer given was that in the first place, not being a genius it would have been impossible to be expert in all the sciences investigating the role which water played in the religion, knowledge and practices of the ancients. Thus, one has to be content with the fact that he studied many books on the archaeology and religions of the Middle East, making use of his knowledge of the hydrogeology of arid and semi-arid regions. Thus, if forced to choose between expertise in one of the sciences investigating history or archaeology, or in hydrogeology investigating the impact of climate changes on the water resources and thus history of this semi-arid region, hydrogeology has preference. This science, besides its importance in providing water to the people of the arid and semiarid regions whose survival depends on finding and developing of this resource, enabled the author to investigate the influence of abundance or lack of water on the fate and faith of the ancients.

At the same time reading the results of the archaeological investigations was a good lesson in the history and religions of the people of the Middle East, about their everlasting fight against the desert, which, last but not least, brought a feeling of respect for the wisdom of these people. This wisdom enabled them to survive and create the books and documents to which the reader will be introduced.

Yet, before the reader begins, some words of explanation are needed regarding the dates and ages used in this book. When discussing geological and pre-historical events the B.P. (Before Present) scale, will be used, which will be replaced by the B.C. and A.D scale once historical events are discussed. Yet, in proto-historical periods or in the case of historical events to which a geo-scientific scale was applied, the author decided to give the two dates. Though the addition or reduction of 2000 to turn one date to the other looks trivial, the author knows from experience that the need to shift from one scale to the other distracts the reader.

References

Bar-Matthews M, Ayalon A (2011) Mid-Holocene climate variations revealed by high-resolution spelepthem records from Soreq Cave, Israel and their correlation with cultural changes. Holocene 21(1):163–171

Bar-Matthews M, Ayalon A et al (1997) Late Quaternary Paleoclimate in the Eastern Mediterranean region from stable isotope analysis of Speleothems at Soreq Cave, Israel. Quatern Res 47(2):155–168

Bookman (Ken-Tor) R, Enzel Y et al (2004) Late Holocene lake levels of the Dead Sea. Geol Soc Amer Bull 116(5/6):555–571

Frumkin A, Magaritz M, Carmi I, Zak I (1991) The Holocene climatic record of the salt caves of Mount Sedom Israel. Holocene 1(3):191–200

Karlen W (1991) Glacier fluctuations in Scandinavia in temperate palaeohydrology in fluvial processes. In: Starkel L, Gregory KJ, Thornes JB (eds) The temperate zone during the last 15000 years. Wiley, New York

Lemcke G, Sturm M (1997) d18O and trace element as proxy for the reconstruction of climate changes at Lake Van (Turkey): preliminary results. In: Third millennium B.C. climatic change and old world collapse. In: Dalfes H, Nüzhet et al (eds) Series: NATO ASI series (closed), vol 49. Springer Verlag, Berlin

Chapter 2
In the Beginning

Abstract The people of the ancient Middle East took for granted that before creation chaos and darkness ruled. Their story of creation is thus the story of putting order in the dimensions of space and time. As light was a condition for life it was first on the agenda, followed by further steps from putting order to the bodies of water, creating vegetation to ensure food for the world of life. Once observers were created, also the day and night came under order. Life commenced in the hydrosphere and atmosphere followed by the creation of life on land and *Homo sapiens* was the last to appear on the sixth day enabling the creator to rest on Sabbath.

> In the beginning God created the heaven and the earth. And the earth was without form, and void; and darkness was upon the face of the deep. And the Spirit of God moved upon the face of the waters. (Genesis 1: 1–2).

Albeit that these sentences and those which follow cannot be considered by a geologist as an answer to the question: "How did it all begin?", still one must agree that these sentences were in the first place answers to the enquiring human mind. Secondly they show that this mind took it for granted that formlessness, disorder and darkness preceded form, order and light.

The logical human step to enable creation is to bring light to the surroundings and indeed this step follows: ***And God said, Let there be light: and there was light (ditto 1:3). A step which also brings order in the dimension of time: And God called the light Day, and the darkness he called Night.*** (ditto 1:5).

As will be discussed in more detail in the following chapter these ideas of cosmic order were asked and answered by the people of the ancient Middle East and most probably were first chanted and later written by the people of Sumer, located at the separate outlets, at that time, of the two rivers, namely the Euphrates and the Tigris (Hidekkel) to the Persian Gulf. The life of the Sumerians was dependent on the water of these rivers which they saw rising when the rains fell on the mountains to the east, and the sea stretched endlessly to the south. Their explanation for the existence of the various bodies of water was, from their point of view logical: ***And God made the firmament, and***

A. S. Issar, *Strike the Rock and There Shall Come Water*, SpringerBriefs in Geography,
DOI: 10.1007/978-3-319-01937-6_2, © The Author(s) 2014

divided the waters which were under the firmament from the waters which were above the firmament: And it was so. (ditto 1:6).

After order was achieved on the spatial vertical ('Z') coordinate, it was requested on the spatial coordinates ('X, Y') and indeed: ***God said, Let the waters under the heaven be gathered together unto one place, and let the dry land appear: and it was so. And God called the dry land Earth; and the gathering together of the waters called he Seas***: (ditto 1: 9–10).

As the people composing these sentences were not living in the desert, then once there is land an explanation for its cover, i.e. the vegetation, has to follow. The creation of which must be prior to the creation of the living world, which will be fed by it. Thus: ***Let the earth bring forth grass, the herb yielding seed, and the fruit tree yielding fruit after his kind, whose seed is in itself, upon the earth: and it was so.*** (ditto 1:11).

The introduction of grasses, seeds, trees and fruits reminded the composers of the changes of seasons: the gloomy winter, with its short days bringing rains, and the change to the sunny, long day of summers with ripening cereals and fruits. The composers and chanters of the story of creation then go back to the creation of the heavens and tell us about the role of ***the lights in the firmament of the heaven to divide the day from the night; and let them be for signs, and for seasons, and for days, and years:*** (ditto 1:15).

At this stage life can appear. Though not being acquainted with the science of evolution, the ancients give priority to the creation of life in the hydro-sphere and the atmosphere: ***Let the waters bring forth abundantly the moving creature that hath life, and fowl that may fly above the earth in the open firmament of heaven*** (ditto 1:20).

The intuition of the ancients regarding the appearance of creatures on the chronological tree of life continues. After giving priority on this tree to creation of life in the hydrosphere and atmosphere they allow life on land to appear. First come the creatures which enable human life, i.e. cattle, followed by other terrestrial creatures: ***Let the earth bring forth the living creature after his kind, cattle, and creeping thing, and beast of the earth after his kind: and it was so***. (ditto 1:24).

Now when all was prepared for the *Homo sapiens* to arrive he is created male first and female second ***Let us make man in our image, after our likeness: and let them have dominion over the fish of the sea, and over the fowl of the air, and over the cattle, and over all the earth, and over every creeping thing that creepeth upon the earth*** (ditto 1:26). Is the highlight put on ***man in our image, after our likeness***, is a remnant from the period that the idols of gods were created in the image of men?

Then comes the seventh day, the Jewish Shabbat, the day was blessed and sanctified because it was day on which the Lord rested. The name of this day is connected to the number seven, in Hebrew *sheva*. (In other Semitic languages the number seven is pronounced in a similar way), and to the word *shavo't* namely 'to cease work'. A week in Hebrew is *shavooa*. Another similar word is: *shvooah* i.e. oath. Thus this number has an element of sanctified in its various appearances, which I suggest is connected with the cult of the moon and had a religious meaning even before the time of the Hebrews.

In the forthcoming chapters I will return to this number. In conclusion the ancient Hebrews adopted the seventh day of every week as the holy day although detaching it from the lunar cult.

Chapter 3
When Tiamat and Marduk Ruled the Worlds

Abstract The stories of creation found in the first chapters of the Bible reflect the spiritual effort of intelligent human beings, the Sumerians, who inhabited the lower stretches of valleys of the Tigris and Euphrates, to explain how the world around them was formed. The infinite stretches of the deep sea bordering their land on the south brought them to crown Tiamat as the queen of the abyss and ruler of the world. This belief was passed from one generation to the next and when writing was invented engraved on clay tablets. The neighbors to the north, the Akkadians-Babylonians, claimed that their highest of all Gods Marduk subordinated Tiamat, and was the creator of the world. These stories were heard by the ancient Hebrews who transformed them according to their system of beliefs and language.

And darkness was upon the face of the deep abyss

This verse of the second from the first chapter of the book of Genesis came to the author's mind in the evening when he was flying in a small airplane over the swamps of the Delta of Shat-El-Arab and the flow of its murky water to the Persian Gulf. It was at the beginning of the 1960 s, when he served as a consultant for developing groundwater sources to the government of the Iranian Shah-En-Shah (King of Kings). One of the sites which he investigated was in the vicinity of the small town of Bushier on the coast of the Persian Gulf, where a power station for electricity, was planned to be built. For the construction it was necessary to find a source of fresh water, while seawater was planned to cool the plant and he was asked to locate exploration wells for the fresh water.

After he finished the survey and marked the sites and instructed the local hydro-geologists, who were also his students, he boarded a small plane to return to his apartment in Teheran. Because of the very narrow coastal plain the air strip was short and limited to small airplanes which had to land twice on their way to refuel. This is why the flight route continued along the water border with Iraq and thus above the Persian Gulf and later above the delta of Shat-El-Arab. It was twilight and the plane flew at a low altitude. Looking

A. S. Issar, *Strike the Rock and There Shall Come Water*, SpringerBriefs in Geography, DOI: 10.1007/978-3-319-01937-6_3, © The Author(s) 2014

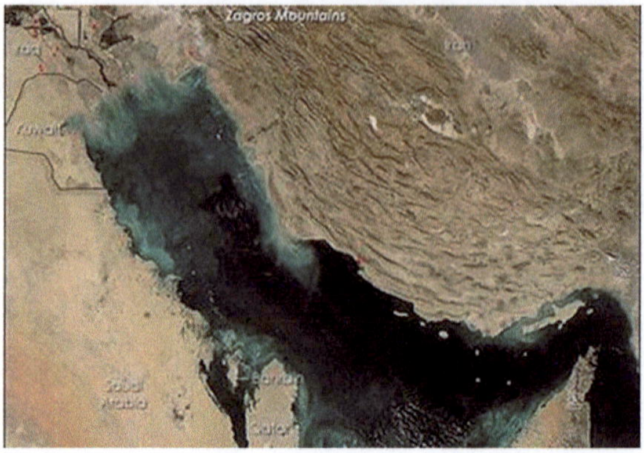

Fig. 3.1 The Persian Gulf and the outflow of the Shat-El-Arab into it (courtesy NASA). The whitish clouds in the water at the outlet of the rivers are a result of the muddy water, which the rivers pour into the gulf

out through the small window he observed that they flew over a thick soup of mud and water, so thick that one could not tell whether it was land or sea. One could not discern any shape or form in these heavy clouds of mud and the term "chaos" aptly described the picture reflected in the water. The evening fell slowly, and with it darkness, and the author remembered the title of the present chapter i.e. ***and darkness was upon the face of the deep abyss*** (Genesis 1:2). (Fig. 3.1).

The night was especially dark, and the only light outside of the airplane was the blinker on the plane's tail, which thickened the darkness. The blackness outside the window and the knowledge that beneath extended a region, neither sea nor land, but a strange mixture of both, deepened the experience of the primordial darkness enveloping our small plane. All these reminded the author the biblical verses of Genesis about the creation of the world, which he had learnt by heart as a young boy and which were discussed in the preceding chapter. The reader can guess that as a geologist he doubts if the biblical story of creation, indeed, describes the history of the world, despite being evidence to the wonder and to the budding scientific thoughts of his ancestors.

At this stage the author is convinced that in the light of various scientific observations we have to accept the theory that our world emerged as a result of the "Big Bang" 13.7 billion years ago. Before the Big Bang the whole universe concentrated in some highly condensed form that exploded and rapidly expanded. In the beginning the universe was a blazing fireball and over time it cooled. The process involved the formation of galaxies like the Milky Way to which our solar system belongs and where planets like our Earth and its–our moon circle the sun.

After the temperature of the globe dropped beneath the boiling point, water began to collect into the lower parts of the globe stretching over its greater part. When the human being acquired the faculty of language he called the big stretches of water, seas and later oceans, while the smaller ones were called lakes. In these bodies of water, before they were named by the hominines, special conditions existed for the emergence of a rather simple form of life. Albeit the fact that the starting point is not yet known, neither the question whether life began on our globe or came from another cosmic body, still the basic knowledge that it all started from the simplest life-form is not debated. This primary form was built from a few molecules that had in them a basic genetic code that guided a line of a series of multiple steps while duplicating the code and transferring it from generation to generation. The preliminary molecule began developing as flora and fauna involving phenomena of random changes in the genetic code which are called mutations. The surrounding natural conditions which changed for better or worse and the competition for the available resources, food and space brought and led processes of natural selection, the significance of which Charles Darwin was the first to publicize.

While Darwin emphasized the evolution of form due to natural selection, in the book *From Primeval Chaos to Infinite Intelligence—Information as a dimension and entropy as a field of force,* Issar and Colodny (1995) claim that the process of evolution of the bio-world is a ceaseless increase in intelligence. In the present author's opinion, this issue can be explained neither by Darwin's theory nor by such neo-Darwinians as Richard Dawkins (1976). The reader interested in the author's views regarding the 'Evolution of Intelligence' is invited to read the above mentioned book and/or his articles (Issar 2008a, b, 2010, 1995, 2012) on this subject, published over the years.

Needless to say that the author's hesitations regarding the scientific truth in the biblical story of creation did not start in his childhood, when he read for the first time the verses in the book of Genesis. Doubts came in his adult life, when he began to be interested in natural science, and especially when he began studying geology, at the Hebrew University. These studies made clear to him the gap between science, based on observations in nature and their interpretation and logical consequences, and the stories of the Bible which are the results not only of observations but also of imagination and the religious beliefs of the ancients. These stories are the spiritual effort of the intelligent human being to explain the world around him.

Still, his field observations while the author was working in the Negev and Sinai deserts, which will be elaborated in the forthcoming chapters, taught him that in the biblical stories one can find information on events which indeed happened. The stories of these events were transferred from one generation to another, interwoven with a religious explanation given by the people who experienced them and by following generations. This view regarding the stories of the ancients influenced his program of work after the author accepted UNESCO's invitation to investigate the impact of global climate changes on the water resources in the different parts of the world. He started by reading the stories and mythologies left by the ancient inhabitants which were available in the countries he visited. From these texts were extracted important information about the impact of events of climate changes on past landscapes and people.

This brought the author to the Babylonian story of the creation, the *Enuma Elish*, a term literally translated as "When on High". He had read the Babylonian-Akkadian version written in the Latin alphabet together with its English translation. The intensity of the expression and the similarity to Hebrew had a strong impression, and led him to translate the ancient epic into Hebrew leaving the words in Akkadian which are similar to those in Hebrew.

The opening lines of the *Enuma Elish,* describing the primordial state of the universe, reminded the author of that night flight over the Persian Gulf. The following are the first lines of the First Tablet of the seven tablets as translated by L.W. King (1902)

When in the height heaven was not named,
And the earth beneath did not yet bear a name,
And the primeval Apsu, who begat them,
And chaos, Tiamat, the mother of them both
Their waters were mingled together,
And no field was formed, no marsh was to be seen;
When of the gods none had been called into being,
And none bore a name, and no destinies were ordained;
And the earth beneath did not yet bear a name,

It is very possible that when the ancient Sumerian poet, living on the northern swampy shore of the Persian Gulf, composed the first version of the story of creation, writing was not yet invented. This poet, who may have been the head of a tribe, or a spiritual leader, recited the verses to his people. They memorized them and recited them to their children, who recited them to their descendents, generation after generation. Many years later, when writing was invented, the verses of creation were engraved on clay tablets. Those were transferred to the neighboring people, the Akkadians-Babylonians, and they added them to their story of the deeds of the most high of all Gods Marduk, the subordinator of Tiamut - Tiamat the former ruler of the world. These, and Marduk's placing the sun and moon in their tracks, and all his other wonderful deeds are engraved on the seven tablets of the *Enuma Elish*.

These wise people, who knew many things, didn't yet know how to fly above the earth and picture the swamps from a bird's eye view. Moreover, at that period the Euphrates and the Tigris were not yet united creating the Shat-El-Arab, and the both rivers flowed to the Persian Gulf separately, but there is no doubt that the wide marshy landscape already existed then at the outlets of the rivers to the sea. The ancient inhabitants, most probably, like the present dwellers of the swamps, sailed on cane rafts in the waterways winding among the multitude of islands created by the mud and the rivers' silt deposits. The land was covered mostly with water and the dryer patches were covered by thickets of canes. Tracts of marshes stretching among the islands swallowed anyone who happened to fall from the raft. To these people it seemed as if a huge monster opened its mouth and swallowed the man. Most probably the ancient Sumerians sailing on the turbid waters of the channels and along the shores of the gulf saw in the muddy water forms of monsters who gazed at them ready to swallow them. Their imaginations, doubtless more lively though less scientific than ours, flying over swamps. They named this ancient monster of the deeps *Tiamat*.

When the stories and beliefs of the ancients passed from mouth to mouth and from one generation to another, they were heard by the ancient Hebrews who transformed it into their system of beliefs and language. Thus Tiamat turned to be *tehom* meaning an abyss, and even *Tehom Raba* namely Great Abyss, while *Tiamat* was forgotten, until the English archaeologist and Assyriologist Leonard William King

(1869–1919) came and deciphered the Babylonian-Akkadian writing and made *Tiamat the Goddess of Abysses* to re-appear suddenly from the abysses of oblivion.

Returning to the flight over the swamps of the Shat-el-Arab and to *Tiamat*, at that time the author already held a Ph.D. in hydrogeology the science of groundwater research. In Hebrew the term groundwater is *mey-tehom* namely water of the abyss. Previous to his studies of hydrogeology, and due to his acquaintance with the dissolution caves in the limestone and dolomite rocks building the mountains of Jerusalem where he was born and raised, the Hebrew term of *Tehom*, led the author to imagine that groundwater flows as subsurface streams in such caves underground. The science of hydrogeology explains the cycle of water in the atmosphere as precipitation and the path the water takes after infiltration through the permeable land cover, whether soil, sand or rock to reach the porous saturated part of the subsurface to form groundwater.

In the story of the deluge, to be dealt with in the next chapter, the Great Abyss is the source of the fountains that spouted forth and flooded the land in addition to the rain water falling from heaven. In contrast to this, in the blessing of Yaakov that wishes for his son Joseph ***Blessings of Heaven from above, and blessings of abyss from beneath*** (Genesis 49: 25), the water which comes from heaven and the fountains which spout from the abyss bring blessings. In the Hebrew version of *Exodus* and the crossing of the Red Sea, also to be discussed later, *abysses* swallowed the Egyptians who were chasing the escaping Hebrews. ***Abysses will cover them they went down in depths like stone*** (Exodus 15, 5). In all these verses and many more, the term "abyss" expresses depths of the underground full of water. This is why, the Babylonian-Akkadian text of *Enuma Elish* describing *Tiamat* as a Goddess that represents an infinite water body, revealed the source of the Hebrew term *tehom* (תהום). Yet, while in the Mesopotamian (Sumerian and the Babylonian-Akkadian) tradition this infinite water body, has infinite depth and is very saline, in the Hebrew tradition it refers also to extreme depth, yet the Hebrew the term abyss-Tehom refers to marine as well as continental depths. The Sumerian tablets tell us that the gods Apsu the god of the fresh water, Tiamat goddess of deep saline sea and Mummu the god who represented the mist or even rain did not like the behavior of the gods who were their descendants, and wanted to destroy them and thereby to destroy the newborn world. This scheme, however, was opposed by Ea-Enki: "He of supreme intelligence, skillful, ingenious, He who knows all things. He put a spell on Apsu who fell into a deep slumber. Then he locked up Mummu and passed a string through his nose and sat holding him by the end of his nose rope".

Evoked by the forces of chaos to avenge her mates Tiamat mobilized all the forces of destruction for vengeance. Enlil, in the Sumerian; version, or later Marduk, the chief god in the Babylonian pantheon, defeated Tiamat with the help of the winds, killed her, and then cut her body in two. One part he placed above earth to create the sky, while he made sure that the water in it would not escape by setting up locks and appointing guards on the locks. The lower part formed the "water of the deep".

This story, which most probably stems from the Sumerian mythology, but was later retold by the Akkadians who introduced archaic Semitic traditions, describes the war between the gods, whether Enlil, Assur or Marduk, as chief gods protecting their cities against the forces of destruction symbolized by the powers of the waters of all types. The fact that the evil gods wishing to bring destruction on their descendants were represented by forces of water is an indication, most probably, of the continuous struggle of the societies worshipping the good gods to help them against the havoc caused by the rivers' floods, the upsurge of fresh or salty groundwater, and the encroachment of the saline water from the sea into the shallow deltas of the rivers the Persian Gulf.

This struggle, which started when man first set foot in the valley of the two rivers in prehistoric times, and continues to the present day, was symbolized by the ancients as a war between the destructive and protecting gods. The first were also symbolized by serpents and dragons.

The echo of this symbolic fight later saw many variations and has been told and retold by many generations, each claiming their protector to be the hero who killed and put the spell on the forces of evil. The struggle between the hero-god patron and protector of human society, and the forces of destruction, symbolized by Tiamat knew many incarnations either by God killing Pharaoh, King of Egypt, the great Tanin dragon that lieth in the midst of his rivers (Ezekiel 29:3) or the legend of St. George killing the dragon, which has its origin in Greek and Nordic mythologies, and its offspring are found in folklores of many nations. In Chinese folklore the dragon represents positive aspects of life.

The question is: Where did this basic idea of the water as a primeval force of destruction tamed by a divine constructive force originate? Was it in the fertile valleys of Mesopotamia or Egypt, or in the plains and valleys of Canaan, Anatolia or Iran? Surveying the hydrological conditions of Mesopotamia one is inclined to see in this part of the Levant the origin of the theme of war between the powers of goodness, those ordained to supply shelter, food and thus life, against the evil powers: famine, destruction and death. If indeed the origin was in the plain of Mesopotamia most probably it started with the Sumerian people. The Akkadians, arriving later to this plain with their stories of creation, adopted the principal motifs of the Sumerian mythologies with some modifications, to better fit the new environment. In Mesopotamia, their new homeland, the water, when uncontrolled as either floods or saline water encroaching on the fields, indeed meant destruction. This brought about the organization of a social system tasked with building canals, embankments, and drainage channels, as well as their constant maintenance. This organized society was governed by a king who received his power from the divine protector of the city which afterwards expanded to become a state.

The story of Marduk overpowering Tiamat, Apsu and Mummu, as told in the *Enuma Elish* was recited at the festivities of the New Year in Babylon or Bab Illim (*Bab* in Akkadian means gate, while *Ilim* means gods). During these festivities, celebrated at the beginning of spring, namely the first days of Nissan the month of spring (the Jews in exile in Babylonia adopted the local annual calendar, including the names of the months, and celebrated Passover, the feast of spring on the fifteenth namely mid-Nissan). During this month, the first month of the Babylonian calendar, the king went through a ceremony of dethronement followed by re-coronation. These festivities of the New Year were the time to repent and request mercy in order to protect man from the destructive powers of nature. These New Year festivities, which as mentioned, were celebrated at the beginning of the spring marked the end of the winter. During the spring the warmer climate brings the flourishing of vegetation, of the grasses in the pasture lands, the blossom of the trees, and the birth of young.

The literate history of the Sumerian mythology began about 5,000 years ago. As mentioned the Babylonian-Akkadian tradition of the victory of Marduk over Tiamat is based on the Sumerian mythology. It is quite obvious that the ancient traditions came down from earlier times when stories were passed verbally from one generation to the next. The mentioning of water, as well as the many symbols connected with it, points to the fact that the traditions grew in a country which was endangered by, as well as benefiting from water. As the Sumerian civilization had had its roots in the Mesopotamian plain since prehistoric times, it is logical to look for the roots of these mythological traditions in the interaction between the natural environment and man in this area. In Sumerian mythology the fight between the benevolent and the evil gods also contains the subject of primeval waters ruled by the evil god. In one of these stories Kur, the god of the netherworld, who controls the Primeval Waters is involved in a foul deed against the sky goddess, Ereshkigal. Enki the water god, sets out in a boat to attack Kur who fights back by sending the primeval water to sink Enki's boat. The tablet does not tell the end of the battle, but from other sources it is known that Enki, the good god of water, triumphed.

> This tablet starts with the following lines:
> After heaven had been moved away from earth
> After earth had been separated from heaven
> After the name of man had been fixed… (Kramer 1961)

Thus, once again, one finds the tradition of creation by the separation of Heaven from Earth connected with the story of a fight against the Primeval Waters.

The story of Tiamat and Marduk and also that of God, dividing the land into the sky or upper water and sea or lower water, can thus be seen to have developed not just from one ancient story, but from a cumulative process of experiences starting in prehistoric times when man had just begun to penetrate the plains of the Euphrates and the Tigris and continuing through the period when history started to take shape.

The picture which evolved before the eyes of the intelligent Homo when he first came into the plain of the Euphrates and Tigris was that of a land coming out of chaos into some kind of pattern–a marsh of saline and fresh water covered by mists which does not enable the boundary between sky and land or water to be seen. (Fig. 3.2).

Another possibility is that the fertile land emerging from the domination of the waters was a land of clash between two races of people: on the one hand, the

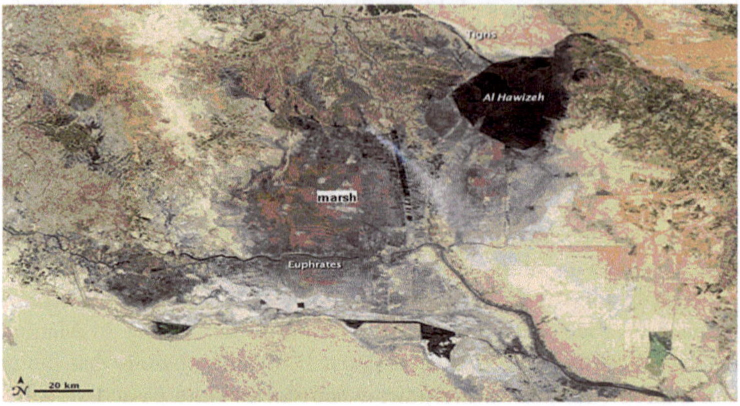

Fig. 3.2 Satellite image of the marshes along the Euphrates (courtesy of NASA)

people coming down from the highlands where either overpopulation or reduction in precipitation forced part of the population to immigrate, and on the other hand, people coming from the fertile plains of the Levant, which was slowly turning into deserts. The story of the fight for life or death between Cain the land tiller, and Abel the pastor, may be an echo of this clash, which ended by these lands adopting the material culture of Cain while they also absorbed the beliefs and religions of the pastoral people.

Thus, the story of the creation of the land out of the primeval water and chaos was passed on from generation to generation. Returning to these stories of creation once the picture of the natural environment has become clear, one cannot but admire the verbal power of the ancients in conveying to their descendants the natural processes which they experienced. Starting with "*When in the height heaven was not named*" etc. as cited above.

Granted that these myths deliver to us the beliefs and fears of the settlers of the plains of Mesopotamia as well as their explanations of the wonders of nature they experienced, the question remains, can we correlate these stories with specific historical events, or are the myths just some kind of a mixture of many events that have lost any traceable connection with what really happened?

References

Dawkins R (1976) The Selfish Gene. Oxford University Press, New York City

Issar AS, Colodny RG (1995) From primeval chaos to infinite intelligence–Information as a dimension and entropy as a field of force. Ashgate-Avebury, UK

Issar AS (1995) Impacts of climate variations on water management and related socio-economic systems. Technical documents in hydrology, IHP-UNESCO, Paris, Sc.95/Ws.26, pp 97

Issar AS (2008) Evolution of intelligence in the bio-world divine action and natural selection. In: Questions of science and faith in biological evolution. Gordon R, Seckbach J (Eds.) World Scientific Publishing, Singapore

Isasr AS (2008) The evolution of complex information systems as movement against the pull of entropy, in a space-time-information continuum, *Physicaplus* No. 10. (Online Magazine of the Israel Physical Society IPS.)

Issar AS, Sorek S (2010) The possible dimension, additional to space-time, which physicists ignore. J Mod Phys 1:70–76

Issar AS. (2012) The continuous increase in the complexity of the designed structures of the universe is described as movement against infinite entropy. In: Swan L, Gordon R, et al (Eds) Origin(s) of Design in Nature, (Cellular Origin, Life in Extreme Habitats and Astrobiology, Vol. 23) Springer, Dordrecht, NL

King LW (ed.) (1902) *The Seven Tables of Creation*. vol.1, Luzac, London, pp 274

Kramer SN (1961) *Sumerian Mythology*. Harper and Row, N.Y. pp 130

Chapter 4
Tiamat and Marduk Depart, Enter Adam and Eve the Tool Makers

Abstract Flint tools, found in northern Israel tell that hominids reached the Middle East about two and half million years ago. Other sites of later dates contain burnt wood, which may be the earliest evidence for the controlled use of fire by hominids. The migrations from Africa were during glacial periods when the Arabian-Saharan desert belt enjoyed a humid climate. The cold glacial climates, which made the deserts green and passable, brought also a decrease of amounts of the rains over the tropics and subtropics. This caused a decrease of food sources and thus migration of animals and hominids and later also the *Homo sapiens* who left eastern Africa around 100,000 years ago and migrated to Asia. His remains were found in caves in Israel. Thus climate changes, which affected the natural environment were significant in determining the history of the Middle East.

> And the LORD God formed man of the dust of the ground, and breathed into his nostrils the breath of life; and man became a living soul. (Genesis 2: 7)
> And the LORD God said, It is not good that the man should be alone; I will make him a help meet for him. (Genesis 2: 18)

Although the oldest written documents come from Sumer, archaeological investigations show that civilized societies existed in the Fertile Crescent before man settled in the Mesopotamian plain. Man was in the Fertile Crescent already at the dawn of the history of mankind; as a matter of fact, he was there even before he could be called man. Some type of a mythology might even have developed which man brought with him to Mesopotamia as a verbal tradition before he knew how to put it in writing. In the following chapter the foot marks of the pre-*Homo sapiens* will be examined.

The choice of prehistoric hominid to settle in this region, which is semi-arid due to its position on the margin of the global arid belts, is an obvious indication that this border was not static and that this semi-arid zone was more humid during certain periods. Indeed, one can find signs of colonies of prehistoric hominids in the middle of the desert near sites which were once either swamps, lakes or springs, and which dried up when arid conditions prevailed. At these locations were found flint tools, bones, and teeth of animals killed by the settlers.

A. S. Issar, *Strike the Rock and There Shall Come Water*, SpringerBriefs in Geography,
DOI: 10.1007/978-3-319-01937-6_4, © The Author(s) 2014

Sometimes even hominids bones have been found. Thus, throughout pre-history and history climatic changes which brought humid conditions were also periods of profusion of vegetation and life in these regions.

The earliest evidence for hominids' intrusion into the Middle East were flint tools were found at a site in the Galilee, Israel, covered by a basalt flow dated to around 2.4 million years (Ronen 1991). This site is, for the time being, the only place in this region where such an ancient culture was found (*Homo habilis*?).

Another site with evidence of the presence of hominids in the Middle East is in the Jordan Valley about 3 kms south of the Sea of Galilee at Tel Ubeidiya. It was discovered in 1959 by a member of a nearby Kibbutz, who notified the archaeologist Moses Stekelis. A team consisting of the zoologist Georg Haas, the geologists, my teacher Leo Picard and his staff, started the excavations which were continued by the archaeologists Ofer Bar-Yoseph and Naama Goren-Inbar (Picard Baida 1966). These excavations proved this site to be the richest with Lower Paleolithic tools. On the basis of geological and faunal correlations the age of the layers in which the tools were found was estimated to be Lower Pleistocene about 1.4–0.7 million years ago. The toolmakers were living on the shore of a marsh or shallow lake and the abundance of fauna with northern affinities indicates a cold climate. The 150-m Ubeidiya Formation has been the subject of systematic fieldwork for several decades. Some small fragments of hominids' cranial vault bone and tooth were found. Initial age estimates suggested ages of 0.8–1.0 million years based on pollen correlations (Horowitz 1989).

Another place with evidence for early migration of hominids from Africa is at 'Erq el-Ahmar in the northern part of the Jordan valley, Israel, where tools of Lower Palaeolithic Oldowan type were found. New paleo-magnetic results from this site indicate that these sediments were deposited between ca. 1.7 and 2.0 million years ago (Ron and Levi 2001).

At Gesher Benot Ya'aqov, in the Jordan valley north of the Sea of Galilee another site with Paleolithic -Acheulian tools dating about 780,000 years ago and animal bones have been unearthed. The layers indicate that a lake existed in this place on the shores of which dwelt groups of hunter-gatherers. After the lake vanished the site was still traversed by the perennial Jordan River, which during winter months may overflow its banks and flood the area and thus the clayey deposits of the ancient lake remained waterlogged, preserving the remains of the plants living around the lake and over its drainage basin. A few of the pieces of wood show signs of artificial modification—making them the oldest wooden artifacts yet discovered. The remains of the natural vegetation enable the botanists to reconstruct the ancient natural assemblage of plants which existed in this region before the changes caused by human agricultural activity. Concentrations of burnt wood may be evidence of use of fire by the ancient inhabitants of Gesher Benot Ya'akov, which is the earliest evidence for the controlled use of fire by hominids (Goren-Inbar et al. 2002, Alperson and Goren-Inbar 2010).

An assemblage of very primitive flint tools was found in lake deposits forming the banks of Wadi Zihor in the arid southern Negev desert, Israel. These lake sediments are of Lower Pleistocene age as they over deposits of the Arava Formation of

Pliocene age. The Pleistocene lake sediments containing the tools intercalate with horizons of reddish paleo-soils. These layers indicate that the climate during the period of deposition was humid enough to support the formation of a lake, supporting various forms of life including hominids. The fossil assemblage in the freshwater limestone deposits of the lake points to the existence of a permanent freshwater body (less than 1,000 mg Cl/L). The lake existed for about hundred thousand years. The prehistoric artifacts found near the ancient lake belonged to two groups. The first, of which may be contemporaneous with the lake, while the second is found mainly on the younger terraces. The more ancient of the two assemblages of the tools resembles that of Ubeidiya and is composed mainly of heavy primitive choppers. The age is thus around 1.4 million years. The younger assemblage is of upper Acheulean type and is about 250,000 years of age. The concentration of the artifacts is evidence that the Nahal Zihor area provided a favorable environment for hominids during the early and middle Pleistocene. The climatic conditions which enabled the development of a permanent lake in the early Pleistocene did not return again in this site (Ginat et al. 1996, 2003, Issar et al. 2012).

Another site containing choppers was found by the present author further north in the Negev highlands. At present the area is a desert playa with an average annual precipitation of about 80 mm. The tools were found in layers of conglomerate inter-fingering with reddish clays and layers of silt deposited either in a shallow lake or a marsh. The tools are early Acheulean and date from about 1.0 million years ago. The upper layer built of silt and clay inter-fingering with pebbles, of mid Pleistocene age, contains upper Acheulean tools dating from between 500,000 and 100, 000 years (Issar et al. 1984).

Which climate system is responsible for the creation of Lake Zihor and favorable human and wild life habitations during part of the Middle and Upper Pleistocene? Was it a warm climate, during an Inter-Glacial Period which caused the northward migration of the subtropical rain system of the western Indian Ocean or a Glacial Period which caused the southward migration of the polar jet stream, thus causing the cyclonic cold air mass to bring rain to the desert areas of North Africa and the Middle East, including the Negev? The answer is that it was due to a Glacial Period. This can be learnt from the fact that the same Lower Paleolithic stone artifacts were found in other, more northern arid parts of the Middle East in the vicinity of paleo-lakes (Biglari et al. 2004). It is difficult to accept the explanation that rainstorms from the Indian Ocean traveled to this northern site. Additional evidence, for the northern origin of the storms at this time for the upper part of the Quaternary, comes from the dripstones in the karstic caves of Tzavoa and Dragot in the northern part of the Negev desert in southern Israel. In these caves stalagmites and stalactites bear evidence for a humid climate from ca. 200,000 to 13,000 years ago. The isotopic analysis of these dripstones tells that "most of the rainfall that reached Tzavoa cave in the past originated from northeastern Mediterranean Sea moisture." (Vaks et al. 2006)

During glacial periods the Negev Desert enjoyed a humid climate; this answers the question why hominids, with their still rudimentary intelligence, simple social organization, and equipped with primitive stone tools, dared to traverse the desert barrier. The climate change was a severe, on one hand decreasing the rain systems greening the tropics and subtropics causing food sources to decrease and on the other hand greening the desert and making it passable.

This also explains the migrations of the *Homo sapiens* who left eastern Africa some time before 100,000 years ago and whose remains were found in caves in Israel. Indeed clays deposited by rivers denoting a major regression of the sea and heavy rains causing floods carrying heavy loads of mud were found in a subsurface section drilled along the shoreline of Israel and dated to be some time later than 180,000 years ago (Issar 1968, Porat et al. 2003). The climate during the time of the stone culture that followed (Levallois-Mousterian culture) was most probably also humid and cold. Layers of this period were found in the caves of Kebara on Mount Carmel and Hayonim Cave in the Galilee (Ronen 1982). This was the period when modern humans spread out of Africa and expanded through the Middle East into Europe and possibly into central and northern Asia. The dating of the charcoal remains from Kebara Cave, Mt. Carmel have shown the age of the layers to range between 45,200 ± 700 ^{14}C years B.P. and ca. 43,600 ± 600 ^{14}C years B.P (Rebollo et al. 2011). This was the period of the Last Glacial which occurred from ca. 80,000 B.P. to about 14,000 B.P. The humid conditions which characterized this period in the Fertile Crescent terminated at about 10,000 years B.P. The climatic conditions during these periods need special elaboration as they serve as the key for understanding the climatic changes during historical times.

The decoding of the paleo-climatic picture started with the observation that the age of the fossil groundwater under the Sinai and Negev is that of the Last Glacial period and that the isotope content of this water (Oxygen 18 to Hydrogen 2 [Deuterium]) is lighter than that of the water of the present. Later, another investigation showed that rain storms which come to the Negev over northern Africa have an isotopic composition similar to that of the fossil water (Issar 1985). This type of rain storms was preceded in many instances by dust storms. This coincidence led the author to suggest the conceptual climatic model, already mentioned, which posits that during glacial times the desert and the rain storm belts shifted to the south, causing heavy dust storms, the dust being precipitated by the rain. This explains the formation of the loess layers peculiar to the period of the Last Glacial and along the desert borders (Issar and Bruins 1983). It also explains the encroachment of sands over the loess at the end of the Pleistocene and beginning of the Holocene, as the shift of the belts northward during this period brought Ethiopia under the monsoonal regime, which increased the supply of floods and sediment load through the Nile to the sea and from there to the coasts of Sinai and Israel (Issar et al. 1987).

The end of the Mousterian stone tools cultural period also witnessed the disappearance of Neanderthal man. Did he evolve into a more successful hominid or a fitter descendant, namely *Homo sapiens* or was he exterminated by another race which evolved somewhere else, then invaded the region once dominated by Neanderthal man? To this question there exists no answer yet. What we know for sure is that the Upper Paleolithic man was not different in his skeletal characteristics from any human type of the present, although slightly different racial characteristics began to evolve. Evolution since this period has been mainly cultural. At about 14,000 B.P. the process of de-glaciation took place. The climate changed in the Middle East and became drier and warmer. At about 10,000 years B.P. a new period, the Holocene started, the climate was more or less the same as today. At about 9,000 years B.P. a major change took place in the way of life of many human communities. Man started to cultivate the land. How and where this happened will be described in the following paragraphs.

Additional support to the conclusion that humid periods in the fertile Crescent during the Quaternary were synchronous with glacial periods is the fact that during the Last Glacial period from ca. 23 to 18 K years ago a lake filled the Jordan Valley from south of the Sea of Galilee to north of the Sea of Galilee. Also in Wadi Feiran of southernmost Sinai, lake deposits were found dating between 24,000 and 20,000 years ago (Issar and Eckstein 1969). The same range of dates was found also for the fossil water found under the Sinai and Negev deserts (Issar et al. 1972).

In conclusion it seems that the biblical story of the expelling of Adam and Eve from the Garden of Eden fits rather nicely, albeit not precisely to the archaeological story of the exodus of the hominids from the paradise of the tropics to semi arid plains of the Middle East. This happened during the glacial events, when the monsoon and tropical rain systems weakened. As a result, the optimal environmental conditions that existed until then in tropical and subtropical Africa deteriorated. The same global cooling caused the westerlies, cyclonic belt from the Atlantic Ocean to move southward, making the Saharo-Arabian landmass more hospitable and the Sinai and Negev Deserts passable. These climatic and environmental changes encouraged the first hominids and later the *Homo sapiens*, to migrate northward, to expand their territories of hunting and foraging and to cross the once arid areas to reach the more humid lands of the Fertile Crescent, thus opening the road to all of Asia and southern Europe.

It can thus be concluded that even in the earliest history of mankind, climate changes affecting the natural environment of the Middle East were significant in determining the fate of humanity. This repeated itself during the other glacial periods.

References

Alperson N, Goren-Inbar N (2010) The Acheulian site of Gesher Benot Ya'aqov. Ancient Flames and Controlled Use of Fire. Springer Dordrecht, vol II

Biglari F, Heydari S et al (2004) Ganj Par: The first evidence for Lower Paleolithic occupation in the Southern Caspian Basin, Iran. *Antiquity* 78 No. 302, Project Gallery, 5 pp

Ginat H, Zilberman E et al (1996) New findings of Early Pleistocene lake deposits in Nahal Zihor, southern Negev. *Isr. Geol. Surv*. Curr Res. 10:55–61

Ginat H, Zilberman E et al (2003) Early Pleistocene fresh water lake deposits in the Nahal Zihor, Southern Negev, Israel. Quatern Res 59:445–458

Goren-Inbar N, Werker E et al (2002) The Acheulian site of Gesher Benot Ya'akov, Israel, vol I, The Wood Assemblage. Oxbow Books, Oxford

Horowitz A (1989) Prehistoric cultures of Israel: correlation with the oxygen isotope scale. BAR International Series 497. Oxford
Issar AS, Bruins H (1983) Special climatological conditions in the deserts of Sinai and the Negev during the Upper Pleistocene. Paleogeog. Paleoclimat. Paleoecol. 3(43):63–72
Issar A, Bein A et al (1972) On the ancient water of the Upper Nubian Sandstone aquifer in central Sinai and southern Israel. J Hydrol 17:353–374
Issar A, Eckstein Y (1969) The lacustrine beds of Wadi Feiran, Sinai: their origin and significance. Isr J Earth Sci 18:21–27
Issar A, Karnieli A et al (1984) The quaternary geology and hydrology of Sede Zin. Negev, Israel. Isr J Earth Sci 33:34–42
Issar A, Tsoar H et al (1987) A Paleoclimatic Model to Explain Depositional Environments during Late Pleistocene in the Negev. In: Berkofsky L, Wurtele MG (Eds.) Progress in Desert Research. Rowman & Littlefield, Lanham
Issar A (1968) Geology of the central coastal plain of Israel. Isr J Earth-Sci 17:16–29
Issar AS (1985) Fossil water; under the Sinai; Negev Peninsula. Sci. Am. 253(I):104–112
Issar AS, Ginat H et al (2012) Shifts from deserted to inhabited terrain in the arid part of the Middle East, a function of climate changes. J Arid Environ 86:5–11
Picard L, Baida U (1966) Geological Report on the Lower Pleistocene. Deposits of the Ubeidiya Excavations. Nat Acad Sci Jerus 39(4):1–16
Porat N, Avital A et al (2003) Chronology of upper quaternary offshore successions from the southeastern Mediterranean Sea, Israel. Quatern Sci Revs 22(10):1191–1199
Rebollo NR, Weiner S et al (2011) New radiocarbon dating of the transition from the Middle to the Upper Paleolithic in Kebara Cave. Isr J Archaeol Sci 38 (9):2424–2433
Ron H, Levi S (2001) When did hominins first leave Africa? New high-resolution magnetostratigraphy from the Erk-el-Ahmar Formation, Israel. Geology 29(10):887–890
Ronen A (1991) The Yiron-gravel lithic assemblage artifacts older than 2.4 MY. Isr Aechaeol Korrespodenzbl 21:159–164
Ronen A (ed) 1982 The Transition from the Lower to the Middle Paleolithic and the Origin of Modern Man. BAR International Series. 151 pp
Vaks A, Bar-Matthews M et al (2006) Paleoclimate and location of the border between Mediterranean climate region and the Saharo–Arabian Desert as revealed by speleothems from the northern Negev Desert, Israel. Earth Planet Sci Lett 249:384

Chapter 5
Adam and Eve Depart, Enter Cain the Planter and Abel the Shepherd

Abstract In the Middle East, a semi-arid and partly arid region, survival was endangered by either a shortage of water or a surplus. This led the inhabitants to invest mind and body in various water works to regulate the sources of water. To execute these works the society had to be organized in a certain hierarchy. The projects regulating flows most probably started in small mountain valleys and spread to the wider plains in order to regulate bigger flows. This enabled the development of irrigated agriculture, essential in an arid region. This was the case in Jericho, where the remains of an urban community are found which practiced agriculture at the early dawn of civilization. After several centuries the irrigation systems were expanded to the gigantic scale exercised in valleys of the big rivers. On the border of the irrigated valleys two other systems developed. One was the settled agricultural society dependent on rainwater or winter floods. The second was the nomadic society which had to become accustomed to random fluctuations in the abundance of sources of supply.

> In sorrow shalt thou eat of it all the days of thy life. …In the sweat of thy face shalt thou eat bread. (Genesis 3:17,19).

In humid countries, the availability of water is taken for granted. It is one of the basic items which nature seems to be composed of. However, the sun and the warming fire are things which are not guaranteed daily and thus have to be cared for and thought of perpetually. In arid and semi-arid countries, on the other hand, water is the essence of life which is not guaranteed. Its appearance is irregular and random and when it fails to appear famine and death are the consequences. When the same irregularity and randomness work on the other extreme, it brings an overabundance of this resource leading to destruction and chaos. Thus, it is not surprising that the primeval deities, such as Tiamat, Apsu and Mummu of the societies living in the Mesopotamian plains, already described, were those that symbolized the various forces of water. The victory of another divine force either Marduk, Bell, or Asur, symbolized on the other end, the decline in human's fear of these primeval forces of nature. This happened after humans had succeeded in regulating these forces, either by building dams to tame rivers and floods, build canals and irrigation networks. To execute these projects the organization of human society under a certain hierarchy had to take place.

A. S. Issar, *Strike the Rock and There Shall Come Water*, SpringerBriefs in Geography, DOI: 10.1007/978-3-319-01937-6_5, © The Author(s) 2014

Thus it is quite possible that the shift from a hydraulic Pantheon to warriors and agricultural gods was due to the submission of the aborigine population that worshipped the powers of the local rivers and their floods by people who brought with them their own gods from another environment together with a different organizational system which later enabled them to undertake construction of large projects.

Support for this idea of a new regime and "a new order" can be seen in the fact that Tiamat and her pantheon were always described as destructive forces and as opposing order. As a matter of fact, they are provoked to kill the gods and thus cause the war (at the end of which they were destroyed) because of the "new" ways of the gods who want to install "a new regime" and thus disturb Tiamat's and Apsu's and Mummu's sleep.

From the agricultural point of view this can be interpreted as the opposition of the forces of nature represented by the gods of the people living in peace with them, to the interference of patron deities of newcomers who want to regulate the water resources and benefit from them by using them for irrigation.

Thus, one can assume that the archaic societies in Mesopotamia were not agricultural. This assumption is supported by evolutionary, technological considerations, because in these plains too great a human effort is needed on an extensive geographical scale in order to tame and then benefit from the big rivers. Any small-scale endeavor is doomed from the beginning to destruction by the next year's floods. Thus, just by analyzing the message hidden in the ancient myths, one becomes suspicious that the societies who wrote them and built the big water projects were not the aborigines of these plains but came from somewhere else. Moreover, it is more logical to assume that the first river-regulating projects were initially started in the small mountain valleys by diverting and controlling the flow of water coming from small rivers fed by springs, the perennial flows of which supplied drinking water as well as water for irrigation. Only after learning how to control and benefit from a small but regular and manageable flow, can man venture to measure himself against the giants, the flow of which changes drastically from one season to another.

Where could have such an evolution take place and did these people leave any record of the first steps of their evolution? An answer to this question is found in Jericho in the Jordan Valley where the British archaeologist, Kathleen Kenyon (1957), excavated the remains of an urban community which was found to have practiced agriculture at the early dawn of civilization.

Jericho is situated in the Jordan Rift Valley about 10 km north of the present northern edge of the Dead Sea. The annual precipitation is about 140 mm. The reason for the aridity is that the Jordan Valley, a downthrown rift valley, is in the shadow of the rains which come from the Mediterranean and discharge their water on the mountainous high barrier (Fig. 5.1).

Jericho can claim the title of oldest and topographically lowest town in the world. Its occupation started during the Natufian period, between 11,000 and 12,000 years B.P. and it is located at about 300 m below sea level. The name Jericho comes, most probably, from *yareakh*, which means moon (*yerakh* in biblical Hebrew means month). Was Jericho a place of cult to the Moon God or

Goddess of Fertility? Was the place already called by this name in prehistoric times? This we do not know, but what we do know is that the earliest traces of an urban agricultural community were found in this place. As it is situated in the arid valley of the Jordan River and its perennial water supply is dependent on a large spring, one can claim that, indeed, dependence on irrigated agriculture may have started here.

Jericho is an oasis, the life of which depends on the spring called in Arabic *Ein-el-Sultan* (Spring of the Sultan) and in Hebrew—*Ma'ayan Elisha* (Spring of Elisha). This spring is fed from the rains falling and infiltrating into the permeable limestone mountains of Judea and Samaria (Fig. 5.1). Its flow is about 10 million cubic meters per year. The flow is more or less steady throughout the seasons although some minor fluctuations can be observed. The multi-annual fluctuations are also rather small. The reason for this regularity in such an arid environment is its subsurface large storage of groundwater. This is an outcome of the high permeability of the limestone and dolomite water-bearing layers which are recharged annually by the varying quantities of rainwater falling on the mountainous drainage basin. The stored water is gradually released through the relatively small outlet forming the spring. The regularity of the flow enabled man to build a stable culture based on irrigation in this arid land.

With this background information one can go back about 10,000 years and reconstruct the early stages of irrigated agriculture. This was the time when the influence of the Last Glacial period on the climate of the globe dwindled away and with it the more humid phase over the Levant. The big lake, termed Lisan Lake receded; it had filled the rift valley from the Dead Sea to the Sea of Galilee during

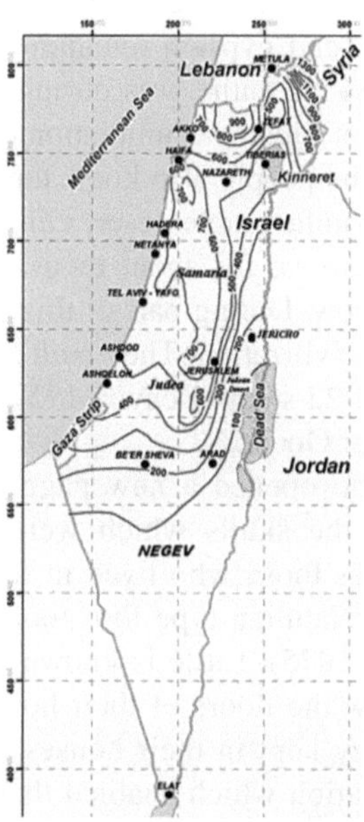

Fig. 5.1 Average precipitation map of Israel, 1961–1990 (Israel Meteorological Service)

most of the period of the Last Glacial. Springs which emerged along the high shores of the lake, at an altitude of about 200 meters below MSL, migrated downward after the receding lake shore. At the same time the climate became more arid. The Homo sapiens who produced the Natufian type flint tools, dwelt on the shores of this lake, and was a hunter, fisherman, and grain gatherer, found conditions of life more and more difficult. It took him about 1000 years to adapt to the new climatic conditions. This adaptation was not just a matter of inventing a new or more sophisticated flint tool. This time the change was in his whole way of life. Man found a way to improve on the natural system of production by the process of seeding and growing wild plants.

This throws light on the finding of remains of grains of domesticated two row hulled barley and wheat in the layers of Jericho dated by carbon 14 to be ca. 9500 years old. The layers belong to the Pre-Pottery Neolithic period when man could not yet bake the most primitive clay vessels but could sow his field, reap the grain, grind it into flour on a specially manufactured grinding stone, and produce dough and maybe even bread.

The climate of the Jordan Valley in the Post-Glacial period was most probably more or less of the same character as today, namely about 120–140 mm of rain with an annual evaporation potential of about 3000 mm. But even if the climate was more humid and precipitation was twice that of today, no cereals could have been grown without irrigation. The proximity of the town to the spring and the finding of the domesticated grains favor the conclusion that people from the Pre-Pottery period did indeed know how to till their land and control the flow of water, diverted from the spring, in such a way that the soil in which the grains were sown would not be flushed away or over-soaked. As the soils of the Jordan Valley contain a high percentage of salt and gypsum remaining from the brackish water of the Lisan Lake which covered the entire area during the Last Glacial period, and as the high ratio of evaporation causes accumulation of salt even of ordinary soils, it seems quite probable that the people also knew the methods of flushing out the salt and its control under the unfavorable desert climate regime. This knowledge, apparently, helped them to survive for about thousand years in this location, as archaeological excavations show. During part of this time, they accomplished one of the wonders of prehistoric civilization. They built a tower 10 m in diameter and 8.5 m high with a staircase of 22 steps (Kenyon 1957). Was the top of the tower a prayer site to the Moon God or Goddess?

Who were the people who opened a new page in the history of mankind? Anthropological studies on the skulls which were found show that the people were of the same stock as those who lived in this region during the former Epipaleolithic period, using Natufian type flint tools, namely, the race or stock called Eurafricans (Mellaart 1975). Little is known about their spiritual beliefs. They buried their dead below the floors of their houses and plastered the skulls of the individuals, which they kept in their houses, most probably as an object of cult. They had an organization which enabled them to mobilize enough manpower to build the tower, which served either as a watch tower or as a cult site, a precursor to the ziggurat of the Mesopotamian plain. They also had commercial relations with distant places like Anatolia from where they obtained obsidian and malachite.

After a period of about thousand years of occupation of the site, the Neolithic Pre-Pottery culture came to an abrupt end. There are signs of flooding on top of the layers from this period. No signs of fire or other destruction could be observed. Was there a climatic change which caused the life at this site to become unbearable, or was it the pressure of enemies who made the tilling of the fields impossible? This is impossible to say. Kathleen Kenyon (op. cit.) maintains that it was indeed a period of floods. As Jericho is situated below the outlet of Wadi el-Kelt, which drains a vast area of the eastern hills of Judea, a more humid period would have caused flooding and even a rise in the water table of Ein-el-Sultan which might have caused the impregnation of the structures and even flooding of the buildings. This hypothesis still needs investigation, as only a rather small section of the layers from this period have been investigated. The humidity profile based on a core drilled at the bottom of Lake Van in Anatolia (Lemcke and Sturm 1997) shows that at ca. 9100 B.P. there was a rise in the humidity of the climate of the Middle East (Fig. 1.5). The people who settled Jericho at about 9000 B.P. had different cultural traditions from the former PPNA (Pre-Pottery Neolithic A) people. The houses they built were better constructed and the type of brick was totally different. The flint tools they used were also different and did not evolve from the PPNA as the latter had evolved from the Natufian. Today, excavations in other sites in the Levant show that these newcomers to Jericho had also previously dwelt in many other sites. They may have re-inhabited the site after a few centuries when the living conditions again became bearable

In the layers of PPNB (Pre-Pottery Neolithic B) as this new layer is termed, a building which contained a standing stone in a niche resembling a Mazaba, namely a cult object in the form of a standing stone, was found. This might have represented the god or goddess of the people of the place. The same can be said about the painted skulls and the three life size figures found in these layers, which comprise over 20 levels of buildings, covering a time span of about 1000 years. The people housed in these buildings continued to live on agriculture. The protein composition of their food was enriched and it contained an additional item of domesticated goat meat, whereas during PPNA the meat of hunted wild gazelle provided the proteins. This finding, namely the domestication of the goat, is important as it demonstrates another economic tool which man acquired and which enabled him to penetrate into the desert areas which have spread since the end of the Last Glacial period. This is because the goat is much more adapted to desert conditions than cattle or sheep. It is thus no wonder that, while the people of Jericho ate mainly goat meat which they most probably herded in the Judean Desert west of their town or along the Arava and the foothills of Trans-Jordan mountains the people of Munhata in the northern part of the Jordan Valley which is more humid, still preferred gazelle and wild pig meat from the swamp of the Huleh, while goat meat formed only a small part of their diet. The domestication of the goat seems also to have held another blessing. It enabled man to cope with a phase of desiccation which seems to have affected this part of the world.

According to the humidity curve of Lake Van at ca. 8300 B.P there is an abrupt rise in the humidity in the Middle East. This may have caused an environmental crisis to the people of Jericho as can seen from the signs of erosion by flowing water observed on the buildings of PPNB. The damages may have been caused by the groundwater table below the town, which began to rise and floods from the nearby wadi to flood the fields and even destroy the houses About 300 years later another crisis occurred, this time by the aridization of the region. The humidity curve of Lake Van Anatolia shows that about 8000 years B.P. an abrupt decrease in the humidity occurred (Fig. 1.5).This climate change may explain the observation of the archaeologists that the PPNB period ends abruptly at about 8000 B.P. That the climate became drier was also observed by the archaeologist Jean Perrot, who arrived at this conclusion on the basis of the reduction in the amount of tree pollen in core samples taken from Lake Huleh in northern Israel (Perrot 1968).

The agricultural systems which began to be practiced in the Jordan Valley, whether exclusively or simultaneously with other sites, predated the agricultural systems in the valleys on the big rivers to the north and the south. Several more centuries were needed before these systems could be expanded to the gigantic scale exercised in the kingdoms of the valleys of the big rivers.

Technically, there must have been a transition phase from the irrigation system with a river as a source. The people of Jericho most probably did not use the water of the Jordan, as this river flows through a deep gorge which cuts into the soft marls deposited by Lake Lisan. If the water from their spring did not suffice, they could have used the springs of the Ein Duyuk and Nuweima, north of their town, or the winter flows of Wadi el-Kelt and its springs. The technical problems involved in diverting water of a rather small spring into fields which lie below its outlet are not too complicated, as large structures and long canals are not needed. On the other hand, when water is taken from a river without a pump being used, a diversion canal with damming structures needs to be constructed upstream and the gradient of the canal has to be controlled in such a way as to avoid destruction of the canal by the flow of water or loss of too much water by infiltration and evaporation due to a very low velocity of the flow in the canal.

A transition stage from spring water irrigation to river water irrigation in a semi-arid zone is found at Çatal Hüyük in the Konya plain on the Anatolian plateau. Settlement of this site began about 8200 B.P., more or less during the period of the late PPNB of Jericho. The variety of cereals grown by irrigation with river water at Çatal Hüyük exceeds that of Jericho. Wheat is the main component. The domestic animal was not the goat but cattle, which supplied the people of Çatal Hüyük with more than 90 percent of their meat (Pritchard 1958). The variety of crops as well as the mixed anthropological types, which were found at Çatal Hüyük speak for the source of seeds, as well as people, from other regions. As was mentioned earlier, the obsidian found at Jericho shows that trading over wide distances was already in practice. It is thus quite possible that the people of Çatal Hüyük obtained some of the seeds and know how from the people of the Jordan Valley.

A striking evolution at Çatal Hüyük had also occurred on the spiritual level. Shrines where a relief of a goddess giving birth to a male lamb and bulls' horn cores set in benches and walls in all probability show that the people of this place worshipped the animal which gave them their meat and milk and most probably also their clothing and work power. Clay and marble figurines of females which were found in the houses were probably used as fertility talismans. Other female statuettes accompanied by broken stalagmites of phallic resemblance can be described as the first pages of a long story on the Goddess of Fertility or the Mother Goddess which was to be transferred from one generation to another. At about 7300 years B.P. the site of Çatal Hüyük was deserted and its people moved to another site in the valley.

A similar system of agriculture based on irrigation from a river, have been found also at Halicar on the Anatolian plateau. The lower levels of habitation at Halicar are contemporaneous with the last levels of Çatal Hüyük but Halicar continued till about 6800 B.P. The abundance of female clay figurines, some of them resembling those of Çatal Hüyük, show that these people also worshiped the Goddess of Fertility.

It is assumed that the skill of irrigating agriculture by diversion of water from a river, which was achieved in Anatolia, migrated to the lower and arid part of the Mesopotamian plain. The fields in the new site were watered apparently from the Tigris, and wheat, barley and flax were grown. The inhabitants kept domesticated goats and sheep and also had domesticated dogs. Female figurines made of baked clay as well as alabaster show that a fertility cult was also practiced by the first irrigation farmers to enter the fertile plain of the two rivers. Was the Mother or Fertility Goddess called Tiamat? Most probably this will never be known, as these people knew only how to express their beliefs verbally and figuratively but not yet literally.

During the time of this early culture (called the Samarra culture) settlements, in other parts of the northern, less arid part of Mesopotamia thrived. These settlements did not yet practice irrigated agriculture, but relied mainly on their herds and the natural vegetative products of the surrounding countryside. These people were using ceramic wares (known also as the Hassuna or Halaf cultures). The transition, however, to agricultural settlements was approaching. But before we concentrate our attention on the settlements and cultures of Mesopotamia we have still to read a few paragraphs on the prehistory of the very early settlements in the western Fertile Crescent.

As has been described, the cradle of agricultural civilization most probably stood in the arid valley of the Jordan outside the walls of the Neolithic town of Jericho. This involved one of the biggest changes in the function of the system called human society, as it became a more deterministic system. How man fared became more a result of his own decisions rather than of outside factors such as migration of herds of wild cattle and abundance of wild plants. True, nature still remained the prime decision-maker. Man acknowledged it and, as an act of submission and fear, was careful to sacrifice to it a share of the fruits and grains of his fields and a few of his domesticated animals. In early times he even offered his own first-born child. Early opportunists and their followers, most probably, soon took advantage of these fears, and demanded unquestioning obedience from the frightened people, to build the early clerical and governmental bureaucracies in order to extract benefits from the working people with a minimum personal effort of their own. But let us not indulge in this story, as this is already another tale of another subsystem with its own rules.

At first, the agricultural societies concentrated in the valleys where they had a perennial source of water to irrigate their lands. Outside the fertile valleys the low latitude semi-arid lands and deserts expanded extensive stretches of land with scant rainfall and hence sparse vegetation and life. As mentioned above, many of the valleys and flood plains themselves received rather poor rains, so the secret of survival of the settler in these valleys was man's ability to harness the water which emerged from the huge springs fed by rains falling on high, distant mountains. At an early stage man learnt to divert this water from its natural channel and irrigate the land he planted. He did this after learning from experience the basic laws governing the flow of water and thus the basic rules of water engineering. As experience continued to accumulate, knowledge continued to grow, and as the organization of human societies encompassed greater numbers of individuals, a combined and organized effort was made to push the irrigated land further into the desolate areas. At the same time some people began to learn how to manage herds of domesticated goats, and later sheep on the arid land, and in this way they may have helped man's society to benefit from the grasses and shrubs which covered the neighboring plains.

On the border of the irrigated valleys two other systems developed. One was the settled agricultural society dependent on rainwater or winter floods. This society can be described as a system that became accustomed to periodic rather than harmonic fluctuations of supply. The second was the nomadic society which had to become accustomed to random fluctuations in the abundance of sources of supply. These people learned to understand the secrets of resource management of the desert by constantly shifting their place of dwelling. It is obvious, that there existed also in-between societies which practiced a mixed way of life, namely were semi-nomadic. Contemporary observations show that as long as the natural resources were abundant the various societies lived in peace and even in some kind of symbiosis.

The periodicity built into the genetic code of the grains, which were the main sources of food, made the societies which were dependent on this food to be observant to the seasonal calendar. The limited quantity of water available for the lands on which these grains were grown brought the settlers to acquire another ability, namely that of measuring quantities, of water and of product. Man had also to learn to measure time in order to know exactly when to apply the water to the fields. At the same time, while making them ready to receive the water, he had to measure the area and afterwards the quantity of water which he applied to the fields. As a result, institutional organizations and conventional metric systems of space and time had to be agreed upon in order to enable numerous consumers to use one source. As already suggested man entered the vast flood plains of

Mesopotamia from valleys fed by springs and small rivers. In the big plains quantities of water and areas of lands were many orders of magnitude larger than in the small valleys. It became necessary to measure large spaces as well as slopes and large volumes of water and afterwards the crops which had to be shared among all the people who had contributed to the big projects. This brought about the invention of scales of measurement able to cope with large areas and quantities, and the mathematics and geometries which enable man to deduce from the small to the large and vice versa.

On the other hand, in the nomadic societies emphasis was put on mobility as a device which enabled man to benefit from a special system of storage characteristic to arid lands. In other words, society had to adapt ways by which to overcome the random negative changes in climate and hence in food. This they did after learning the secret that one could store enough food even in meager years from desert areas if one was able to roam the wide plains and collect the scant vegetation which succeeded in thriving there. The larger the area the more food could be stored. The storage facilities for nomadic societies, replacing the granaries of the farmers, were the fat and meat of the livestock. The number of head in a herd also symbolized abundance, strength and survival. While the number of livestock was small, water could be stored from one dry period to another in cisterns and reservoirs, but when the number increased, the use of other water sources, such as groundwater, had to be introduced. People who dared leave the safe supply of water from the rivers or springs but still did not want to be dependent on the random system of rains in arid lands had to learn the art of digging wells to find and use groundwater.

The gradual development of the methods of excavating deeper wells for tapping groundwater has enabled man to penetrate deeper into the desert and thus be able to utilize the forage which the desert could supply for his livestock. In the regions which were built of impervious rocks it may be assumed that man built cisterns to collect rain water.

The spread of settlements from the Neolithic to the Chalcolithic (Copper-stone) period, at ca. 6500 B.P. in the Middle East began from the lands where springs were abundant, and moved to locations of shallow water tables where groundwater could be found and easily drawn. Only later, during the Bronze to Iron Age (ca. 3300 B.P.) did man move his settlements to areas where enough water could be collected in cisterns from rain. This trend of evolution can be seen quite clearly along the border between the semi-arid and arid zone in the southern part of Israel where, due to special hydro-geological conditions, the groundwater table is near the surface. In such a location, one can even assume that during more humid periods, such as part of the Chalcolithic, springs emerged. But when the climate became dry, man dug the shallow wells in sites where the water could be readily reached, like stream beds.

The best example is the region of the northwestern Negev through which stream beds (wadis) extending from the mountains of Hebron cut into the loess and sandy layers of Pleistocene and Neogene age which overlie chalk layers of Eocene age. In this area a rather large number of settlements of the

Chalcolithic age have been found. The first settlements of early Chalcolithic age, ca. 6000 B.P., were close to the river-bed in places where the water emerges as springs which flow even today. During the later part of this period, settlements moved to the southeast, where the water table is lower. According to the relics left behind, the settlers continued to enjoy a high standard of living, which means they had an abundant supply of food and water. One of the sites at which the relationship between the evolution of society and of the methods of water utilization can be observed is Beer Sheva. Here in particular the evolution of the settlements is especially related to that of the development of the groundwater resources.

The settlement in the vicinity of Beer Sheva investigated by the archaeologist Jean Perrot, (1956 *op. cit.*) started at about 5500 B.P. Subsurface dwellings were excavated by the ancient inhabitants into the loess soil close to the ephemeral river bed. According to the opinion of the present author, this shows that the climate at that time, although a little more humid, was not altogether different from that of the present-day semi-arid climate. The reason for this assumption is that if the precipitation had been much higher, the loess would have been soaked and subsurface dwellings would not have been feasible. Observations by the present author show that while the average annual precipitation at Beer Sheva is about 200 mm, the infiltration of rain water into the loess soil cover is between 100 and 200 mm. Thus if it is assumed that during the Chalcolithic period, which was a cold and a humid period, the rate of precipitation was double than that of the present, then also the soaking of the loess soil was also double i.e. about 400 mm. Moreover, there exists no evidence that the groundwater table in the chalks and gravel below the dwellings did not rise above the level of the river-bed. Thus the artificial caves were dry throughout the year.

The reasons which brought these people to build their houses in the subsurface are not known. It might well be that they came from a region where soft rocks were widespread and where this type of dwelling was the easiest and most natural to construct. If this assumption is correct, one can even theorize that the rocks in the place in which these people first built this type of subsurface dwelling were harder and more stable than the loess soil, as the first dwellings in their new environment had large rectangular rooms of about 3 × 7 m. These rooms collapsed and the people learnt to build small oval-shaped rooms which were more suitable to the geo-mechanical properties of the soil.

The people were farmers as well as herdsman. The large number of sickles and grinding stones, and the granaries show that they knew how to make use of the wide stretches of loess soil and meager rainfall of this region, relative to the more northern parts of the country. The topographical and the hydrological considerations make it quite obvious that their agriculture was not irrigated. The large quantity of churns found in all these sites shows that the milk products from their herds, which mainly consisted of goats, played an important role in their diet and economy. The Chalcolithic sites of Beer Sheva were occupied for a rather short time. Anthropological research shows that the people came from the north. They buried their children under the floors of their houses probably as part of a cult.

An interesting find are the many flat colored pebbles, which were found on the floors of the underground dwellings. These pebbles were grouped in multiples of the number seven.

This brings us back to the meaning of the number seven, which was mentioned in the concluding paragraphs of Chap. 1. The reader is reminded that the reasons given in the Bible to the name of this place namely 'Beer Sheva' (Well–Seven) to commemorate the vow which Abimelech king of Gerar and Abraham made. In Chap. 9 this explanation and the reason for the holiness of this number will be discussed in more detail.

The findings from the Chalcolithic period may point to the fact that the biblical name was a late explanation for an ancient name, adopted, apparently, by the Israelites and explained by them, while blending them with other folklore stories about this region, connected with the rivalry of the shepherds on the scarce water resources of this semi-arid region.

It is beyond the scope of the present chapter to deal with all the occasions mentioned in the Bible, in which the number seven was included. This will be done at each case separately.

Tel Beer Sheva is another archaeological site in the vicinity of the Chalcolithic site discussed in the preceding chapter; it was dated to the Iron Age, with a few Chalcolithic remains and will be discussed when reaching the story of the people of the Iron Age.

A very rich culture of the Chalcolithic period was also discovered at Tuleilat Ghassul, on the eastern side of the Jordan rift valley southeast of Jericho. Today, this site is devoid of any water resource. It is thus quite obvious that the flourishing of the Chalcolithic settlements was due to a more humid climate which prevailed in the Fertile Crescent during this period.

This humid climate also enabled people to live in the Judean Desert west of the Dead Sea (Fig. 5.1). This area is arid due to its location in the shadow of the rainstorms coming from the Mediterranean Sea on the west. Today the Bedouins graze their goats, sheep, and camels in this area only during the winter and spring when grasses are abundant. It is very probable that at the humid periods the vegetation was more abundant. There exists, however, the problem of a perennial supply of water in this area due to the geological and topographic conditions which do not favor the formation of groundwater aquifers at shallow depths. Thus it seems possible that the water was supplied in the same manner as today, namely collection pits, cisterns, or caves excavated into the soft chalk in the vicinity of the riverbeds. This, however, could not supply permanent settlements. These were located in the vicinity of springs or the shallow water table along the peripheries of the desert, either to the west along the watershed or to the east along the shores of the Jordan and the Dead Sea. One of these sites is near the spring of Ein Gedi, where a cult center has been excavated (Mazar 1970). This center is composed of a big yard and a building with an altar on which many burnt bones of livestock were found. A clay statue of a bull carrying two butter churns was also found in this temple. Many remains of the inhabitants of this period were also found in the caves in the canyons leading to the Dead Sea. Of special interest is a treasure

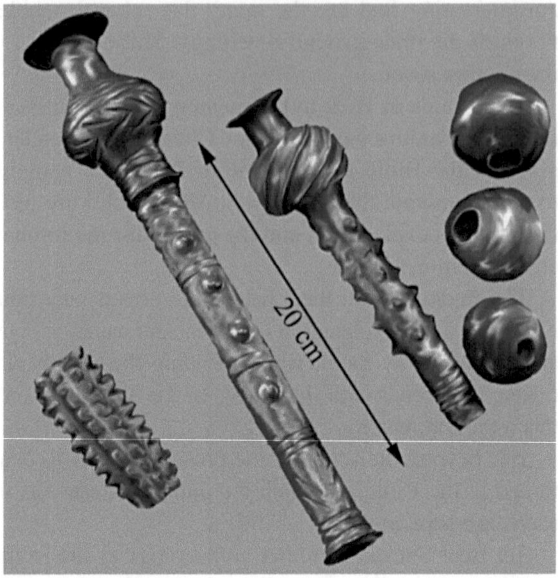

Fig. 5.2 Bronze batons from the treasure cave of Ein Gedi

composed of bronze implements and cult objects that were found in one of these caves. It seems that they were hidden there to protect them from an enemy or when the people had to move from this area (Fig. 5.2) (Bar-Adon 1970).

References

Bar-Adon P (1970) The caves of Nahal Mishmar. In Enc. for Arch. Excav. Isr. Ant. Res. Soc. and Massada, Israel, pp 354–359. (Hebrew)

Kenyon KM (1957) Digging up Jericho. Benn, London

Lemcke G, Sturm M (1997) d18O and trace element as proxy for the reconstruction of climate changes at Lake Van (Turkey): preliminary results. In: Dalfes H, Nüzhet et al (eds) Third millennium B.C. climatic change and old world collapse. Series: NATO ASI Series (closed), vol. 49, Springer-Verlag, Berlin

Mazar B, En-Gedi (1970) In: Enc. for Arch. Excav. Isr. Ant. Res. Soc. and Massada, Israel pp 440–447 (Hebrew)

Mellaart J (1975) The Neolithic of the Near East. Thames and Hudson, London

Perrot J (1968) La Prehistoire Palestinienne, un Supplement au Dictionnaire de la Bible. VIII Letouzey et Ané, Paris

Pritchard JB (ed) (1958) The ancient Near East, an anthology of texts and pictures. Princeton University Press, Princeton

Chapter 6
The Deluge Flooding the Cradle of Civilization

Abstract The story of the Deluge is found in all the sagas of the ancient civilizations of the Middle East, and is an echo of pre- and proto-historical real events. These took place when the fertile valley of the two rivers was settled by an organized agricultural society which harnessed these rivers. When these rivers went out of control it was interpreted as a disturbance in the ordered life of the gods, and as man's failure to meet their needs. To the Hebrews, who inherited this story this catastrophe was interpreted as a punishment for the social corruption ways of human society. The explanation suggested in the present work is that the story of The Deluge is a distant memory of events due to global cooling, which thus caused humid spells in the Middle East. On the basis of paleo-hydrological data, such events occurred at about 7000 B.P., and extend to the period of 6300 B.P., and from 4900 to about 4300 B.P.

> And it came to pass after seven days that the waters of the flood were upon the earth. In the six hundredth year of Noah's life, in the second month, the seventeenth day of the month, the same day were all the fountains of the great deep (Hebrew: tehom raba) broken up, and the windows of heaven were opened. And the rain was upon the earth forty days and forty nights. (Genesis 7:10–12)

The story of the Deluge is a prevailing mythology of the peoples in the Fertile Crescent. In 1827 the members of The Society of Biblical Archaeology in London were astonished to hear from the Assyriologist George Smith (1876) that he had deciphered this story on an ancient clay tablet in the Akkadian language in cuneiform writing. Since then more ancient versions of this story have been deciphered, and today it is quite obvious that it stems from Sumerian origins. Most probably, it passed from one civilization to another, each explaining it in its own way according to its set of beliefs and faiths.

During the years 1927–1929, the archaeologist Sir Leonard Woolley and his team were excavating in Lower Mesopotamia in the graveyard of the city of Ur in Sumer. Woolley believed that this is the city *Ur of the Chaldees* from which *Terah took Abraham his son* etc……*from Ur of the Chaldees, to go into the Land of Canaan* etc. (Genesis 11:31) (Woolley 1929–1954). The archaeologists were excavating

A. S. Issar, *Strike the Rock and There Shall Come Water*, SpringerBriefs in Geography,
DOI: 10.1007/978-3-319-01937-6_6, © The Author(s) 2014

a shaft in a thick layer of rubbish which for centuries had been dumped outside the walls of this ancient city. The shafts went deeper and suddenly the character of the soil changed. Instead of the stratified pottery and rubbish, they were in a layer of "perfectly clean clay uniform throughout, the texture of which showed that it had been laid by water". The workers thought that they had come to the river silt. Woolley, however, felt sure that they were still high above the level of the marsh. After calculating the measurements, he sent the men back to work and, indeed, after excavating through 2 to 3 m, they again found themselves in "layers of rubbish full of stone implements, flint cores from which the implements had been flaked off, and pottery". Sir Leonard concluded that he had found evidence of the biblical Noah's Deluge "which was not universal, but a local disaster confined to the lower valley of the Tigris and Euphrates affecting an area perhaps 400–100 miles across" (1929–1954). This conclusion however, did not convince the other archaeologists who were working in Mesopotamia. They argued that at Eridu only 15 miles from Ur and lying somewhat lower, no trace whatsoever of a flood was found. The sterile clay layers found at other sites, for instance Kish, Uruk and Lagash belonged to other periods such as the Early Dynastic which were later than those found in Ur. The conclusion of most archaeologists was that archaeological excavations in Mesopotamia have afforded no evidence of a regional cataclysmic Deluge.

This conclusion does not conform, however, with the fact that the story of the Deluge is found in all the sagas of the ancient civilizations of Mesopotamia, Babylon and Assyria, not to speak of the story of the Bible. It even goes back to the Sumerian people, who narrated it in a text found at Nippur dating about 3700 years ago. The question remains whether the story conceals an echo of a catastrophe which happened to the ancient civilizations of Mesopotamia or the Levant, or whether it is purely a myth, namely, a story fabricated on a small incident in a few towns or a group of villages in prehistoric times. This story through the millennia was given more and more of an epic significance by succeeding generations, each adding its own adventures to the original story. The history of Mesopotamia as a region in which floods occurred from time to time, causing local damages, gave credibility to the story.

The question whether there was a Deluge or not has interested, and still interests, many people for various reasons. On one side stand the people for whom the Bible is a holy book and who are searching for evidence which will confirm what they are already convinced of, namely that every word in the Bible is true. On the other side stand many scientists who would like to prove that the Bible is just another book of myths which tells more about the rich imagination of the ancients than about what happened in historical or even prehistoric times.

The present author suggests another approach which is based on his field observations as a hydro-geologist. These observations brought him to regard the biblical stories as an echo and a possible source of evidence of some distant traumatic experience which the ancestors of our civilization may have experienced. These traumas were a result of either natural or anthropogenic causes. In many cases the cause may have been a complex one, as, for example, a natural event such as a climatic change bringing a wave of cold which drove hordes of people from one region to another, spreading war and destruction, or a warm phase which brought dryness, famine and migration.

There are two main reasons that led the author to adopt this approach. In the first place, the Bible, which forms the foundation of his Jewish cultural heritage, portrays a gradual evolution from mythology to history. As episodes narrated during later periods were found to have their archaeological proofs, there is no reason not to assume that stories from earlier periods had some factual roots, too, at their source. In the second place, it is quite natural and scientific to examine every piece of evidence upon which one can put his hands, whether archaeological, geological or anthropological. No one would argue that the Bible cannot be regarded as a document in which, at least, an echo of the experience of former generations can be found. The condition is that one should not take this evidence as words coming directly from the mouth of the Almighty and thus leaving no place for any skepticism and scientific enquiry.

Returning to the story of the Deluge the two best-known tales are the ones told in the Bible about Noah and the Babylonian story of Gilgamesh (Heidel 1949; Pritchard 1958). The moral is entirely different in each of the two stories. In the Bible, the destruction of mankind by the Deluge came as punishment for their evils and Noah was saved because of his righteousness ***Noah was a just man and perfect in his generations, and Noah walked with God The earth also was corrupt before God, and the earth was filled with violence*** (Genesis 6:9–11). In the Tale of Gilgamesh the destruction came because the gods quarreled, and Utnapishtim was saved and became immortal only because the gods favored him. An earlier yet less known story of the Deluge is found in the story of Atra-Hasis (Lambert and Millard 1969). It is also a Babylonian epic story but it bears more resemblance to the Sumerian version of the flood story where King Ziusudra is the survivor of the flood. In Atra-Hasis version the flood comes as a punishment to mankind who made too much noise and disturbed the sleep of Enlil, the head god.

The examination of these stories on the background of our understanding of the climatology and hydrology of Mesopotamia leads to the observation that the hydrological phenomena described in all the stories of the Deluge are very similar, but at the same time are different from what one would expect in the framework of present climatic circumstances. All of these stories describe the flood as a result of heavy rains and the emergence of water from the subsurface. (The clay tablet, with the beginning of the Sumerian story of King Ziusudra is destroyed). For people living in a country where the main source of water for everything was the big rivers, while rain was normally a marginal phenomenon, this is quite strange. Did the catastrophe happen when these people lived in another environment, or was it the drastic change in the natural circumstances that left its strong impression on the people, to be remembered for many generations? It seems quite probable that if such a flood, caused by heavy rains, had happened in a land where such an event happens rather frequently, such as the monsoonal countries, it would not have left such a deep impression.

The explanation suggested in the present work is that indeed a true yet unusual environmental catastrophe lies at the root of the story of the Deluge. In other words, the story was composed of a few different layers, at the basis of which lies a distant memory of events characterized by a change in the climatic regime. This entailed rains which flooded the plains and at the same time caused a tremendous increase in the flow of the mountain springs feeding the rivers, due to a rise of the groundwater table in the limestone aquifers of the Taurus and Zagros mountains. Only the people who were clever enough to embark on a floating device, either an ark or boat, and steer to the mountains for refuge were saved.

When did this happen? Hallo and Simpson from Yale University, in their book *The Ancient Near East* (1971) concluded that the great Deluge of Mesopotamia was somewhat more localized than the mythological picture, but it was nonetheless an historical event associated with a specific point in time. They suggest that it happened around 2900 B.C. (4900 B.P.). They base this suggestion on the fluviatile (surface flow borne) layers which were found in some of the archaeological excavations in Shuruppak, (at present Tall Fa'rah south-central Iraq). There the flood deposit "while more modest, is definitely of a fluvial character, moreover it intervenes precisely between the Jemdat Nasr and Early Dynastic levels, that is at the very point when the relative chronology of native sources places the flood". Examination of Figs. 1.2 , and 1.5 shows that indeed about 4900 years B.P (2900 B.C.) the curve of precipitation, based upon the analysis of the carbon and oxygen isotopes of the stalagmites of the cave of Soreq discussed above and that of levels of the Dead Sea and the humidity of Lake Van indeed show high rates of precipitation and humidity in the Middle East at ca. 4900–4800 B.P.

The paleo-climatic curves in these figures show high ratios of precipitation and humidity during the periods between 6650 and 6450 years B.P. the transition period between Late Chalcolithic to Early Bronze I and from 4800 to 4700 years B.P., namely from Early Bronze II to Early Bronze III (Bar-Matthews and Ayalon 2011). The depleted ratios of these oxygen 18 and carbon 13 in the sediments of the Sea of Galilee also speak for a cold and thus humid climate in the Fertile Crescent (Stiller et al. 1984). According to the archaeological time table it is the Early Bronze Period, a global cold period and thus humid in the Middle East. A correlation with the curve of the movement of the glaciers in Scandinavia (Fig. 1.5) shows that during this period, which extended until nearly the end of the fourth millennium B.P., namely for about seven centuries, the glaciers advanced (Karlen 1991).

Indeed, the findings of Woolley also confirm that there was not just one short 'Deluge' as described in the ancient sagas, but there were many more floods in the land of the two rivers. The stories about these floods piled up in the collective memory of the people of Mesopotamia. At the same time it should be taken into consideration that the abundance of rain and thus floods was not limited to Mesopotamia. This can be seen in the curves of the levels of the Dead Sea (Figs. 1.2 and 1.5) and humidity curve of Lake Van, Anatolia, the deposits of which give the story of climate changes in this region in great detail (Degens et al. 1984; Schoell 1978; Lemcke and Sturm 1997).

For the purpose of understanding the significance of the information derived from the sediments of Lake Van, the special characteristic of the environment of this lake will be drawn based on the studies carried out on this lake since 1978–2012

This lake is situated on the high plateau of eastern Anatolia (Fig. 6.1) and has a surface area of 3,522 km^2. Its maximum depth is 451 m and its length is 130 km. Its water is highly saline (alkaline). As mentioned, the sequence of the lake sediments which is composed of thin layers underlying Lake Van gives excellent information on the sequence of climate changes which affected the environment of the drainage basin of the lake The studies based on samples from core drilling into the bottom of the lake showed that these layers are composed from very thin layers, i.e. lamina, brought by the rivers flowing into the lake. Each lamina has a lower white part, which was deposited during summer, when evaporation was high and an upper black part, which was deposited during winter when clay and organic materials were brought by the floods.

The examinations included analyses of the mineralogical content and the ratios between the various minerals, the annual record of the sedimentation rate, the water content of the sediment, the concentrations of organic carbon and opal. The texture of the sediments provided a record of the lake level history. Dating was based on Carbon 14 and on the counting of the layers. Thus, the age of each layer was calculated, as well as the rate of deposition, namely, the thickness of the layer for each year. Albeit the fact that the number of layers may not represent the true age, as argued by a few scientists, no one argues against the fact that the physical and chemical character of the layers portray changes in the mode of deposition due to climatic changes. The pollen of flowers, trees, and herbs were preserved in the sediments and the counting and ratios of various species could show the type of vegetation which grew in the drainage basin. The information on the history of the lake showed that during the Pleistocene, the level showed a few fluctuations at ca. 19 K years B.P, the lake's level was about 70 meters above its present level. Later around 15 K years B.P. the lake must have been completely dry. At the beginning of the Holocene, starting about 10 K years B.P. the lake

Fig. 6.1 Satellite image of the Anatolian Plateau and Lake Van

began to rise and reached a high stand by about 7500 years B.P., (A fall of the level occurred between 9000 and 8100 years B.P). The level dropped to today's level at about 3000 years B.P. (Landmann and Reimer et al., 1996). These investigations have shown also that during a large part of the Early Bronze period the humidity of Lake Van increased. As the water of this lake comes from the same mountains which supply the Euphrates and Tigris and as the lake is closed, the water and sediments which flow into it remain and accumulate. The other data from the lake's bottom, which express changes of climate, were the rates of sedimentation in the lake. These are proportional to the sediment load of the rivers flowing into the lake which might be proportional to the volumes of water brought by the rivers. Floods in these rivers are an indication of floods in the Euphrates and Tigris as they belong to the same region of drainage. The information one obtains from this diagram is that, although floods of a high discharge are abundant throughout the history of the lake, there are periods in which these floods came in high frequencies. Such periods are most probably those which may have caused the flooding of the plains of Mesopotamia, as they meant periods of higher precipitation rates and higher flow in the river-beds, which together have a cumulative influence. The sedimentation rates per 100 years (Degens et al. 1984) show peak values for the period at about 7000 B.P., and extend to the period of 6300 B.P., namely, the period when the lake began to fill up. Another period of rather high values is that from 4900 to about 4300 B.P., which fits the dates suggested by Hallo and Simpson (1971) for Lake Zeribar in Iran where the pollen analysis shows that, at about 5000 B.P., there is a maximum in the ratio of pollen of oak, which denotes a more humid climate.

Another record from which climatic changes can be inferred is the acidity and the isotopes ratios of oxygen 18–16 in the ice cores of Greenland (Hammer et al. 1980). This information is, however, global and does not indicate how a certain change influenced a local system as do the cores from Lake Van on the floods of the rivers of Mesopotamia. Moreover the exact dates of the various events are questionable; being based on carbon 14 dating; they need verification by more advanced methods of analysis (Hammer et al. 1987). Yet, from the acidity of the ice cores one can learn mainly of strong volcanic explosions, containing gases rich in sulfur. This, as mentioned, was correlated with the oxygen isotopes ratios of the ice which is indicative of colder climates. In the opinion of the scientists who investigated these cores, the volcanic explosions threw a large amount of ash into the atmosphere which caused a reduction in the solar radiation and thus lower temperatures. In these cores, one can find supporting evidence for the triggering of a colder period at ca. 6400 years B.P. and from ca. 5200 to 4700 B.P. and thus supporting evidence for the prehistoric flood envisaged by Hallo and Simpson (1971), while taking into account that the exact chronological correlation may change in the order of magnitude of 100–200 years. The evidence from the drainage basin of the Jordan, which starts from the mountains of the Anti-Lebanon and reaches the Dead Sea, speaks also for a cold and humid period during the Early Bronze Period. As can be seen in Fig. 1.2, the curves of the Dead Sea levels, based on the investigations on the caves of Mount Sodom, a salt plug bordering the Dead Sea (Frumkin et al. 1991) and ancient shore lines the level of the Dead Sea was very high during the Early Bronze period. Evidence from the core samples taken from the Sea of Galilee including vegetation history also supports the conclusion of a cold and humid climate during the same period (Baruch 1986).

In conclusion, it is suggested that the mythological stories of the flood of Mesopotamia and the story of the biblical Deluge are an echo on the Dimension of Information (Issar and Sorek 2010) of pre- and proto-historical real events which occurred during and after the time man settled in the fertile valley of the two rivers. The main impact on his heritage was due to events which occurred after man learnt to harness the water of the big rivers for his benefit and learnt about the sources of these rivers from big caverns in the mountains, as well as of the existence of underground waters in the plains. Then suddenly, these waters went out of control and brought destruction to his ordered and organized world. As a result, his whole traditional world view was disturbed, and he looked for an explanation for this disturbance. In the pagan world the forces of nature were represented by a multitude of deities ruled by primeval forces of chaos. Thus such a catastrophe was interpreted as a disturbance in the ordered life of his gods, by a war or just by a quarrel between them, in other cases as man's failure to meet their desires for food and presents. To the Hebrews justice commanded by their abstract Lord was the foundation of order in the world and such a catastrophe was interpreted as a punishment for the corrupt ways of mankind. In addition to these conclusions regarding the correlation between the stories of the ancients and climatic events, it is the opinion of the author that through a careful study of the biblical text and its comparison with the ancient Babylonian texts, and keeping in mind what is known of the hydrological environment of Mesopotamia, one can learn quite a lot about the world view of these ancient societies. One can see that, indeed, many of the archaic traditions originated in Mesopotamia and had been transferred from one environment to the other, possibly verbally. The Hebrew people who afterwards put them into writing did not know the exact meaning of all the words in the archaic Akkadian not to speak the Sumerian text. For example, in the Akkadian version of Atra-Hasis, which might be the nearest to the Sumerian original, the god orders Atra-Hasis to use his hut built of canes ("kanim" in Akkadian and Hebrew) and pitch it (put *cofer*). The later writers of the biblical text and following them the translators to English wrote: ***Make thee an ark of gopher wood; rooms shalt thou make in the ark, and shalt pitch it within and without with pitch*** (Genesis 6:14). Thus although the ancient writers of the Bible used words identical to the Akkadian for pitching, changed the type of the wood from canes to gofer which is very similar to *cofer,* namely pitch in the original text, and instead of *kanim* (canes) used the word *kinim* for which the nearest Hebrew meaning is birds' nests and translated in the English version of the Bible to rooms.

Another interesting observation is related to the term *mayanot tehom raba*, translated as ***the fountains of the great deep*** (Genesis 1:2) as cited from Genesis at the beginning of the present chapter. Thus despite the fact that in the story of the Deluge, and, previously, in the biblical story of Creation, the term *tehom* had already lost its direct association with Tiamat and the fight between Marduk and Tiamat, the term *tehom* is used to describe the primeval and lower deep water. Also, although the story of Marduk dividing Tiamat in two was forgotten, the same conceptual model of two bodies of water still reoccurs in all biblical descriptions of the body of water of the sky and that of the abyss.

Another remnant of the story of Tiamat in the Bible of the Hebrews is *ayin* namely eye, which in Hebrew and Arabic means a fountain as well as a spring of water. (It is interesting to note that the Persian [an Indo-Aryan language] term for spring is *cheshme*, which is related to *chash*, which means eye.). It seems that the archaic pictorial concept, and thus the explanation as to the appearance of springs, was that they were the eyes of Tiamat. The word *ma'ayan*, which also means a spring of water, might thus be an ancient complex form of the words *mei'ayin*, namely water of the eye, most probably the eye of Tiamat. That the association of springs is with the eyes of Tiamat seems quite natural when one takes into consideration that the two large rivers, the Tigris and the Euphrates emerge from large springs, most of them karst springs (namely limestone solution outlets) along the foothills of the Zagros mountains on the border between Iran, Turkey and Iraq. These springs in many instances form large circular pools of water out of which a stream flows (Fig. 6.2). In many cases the springs emerge from big caverns.

In the British Museum one can see on an Assyrian stele a pictorial description of Shalmaneser III's visit to such a cave at the source of the Tigris. The relief shows cattle sacrificed at the source, while workmen are busy cutting commemorative reliefs on the rock (Reade 1983).

When severe rain storms occur on the mountains, the springs respond with a strong outflow, causing the rivers to swell and overflow the plains. The people lived on the plains and their existence depended on the river water they saw the rising of the springs as resulting from the will of the gods, for better and for worse. Thus the killing of Tiamat and the closing her outlets namely her eyes was grasped as the victory of the good gods or god upon the evil ones. Therefore it is suggested that the Hebrew phrase *mei tehom* or fountains of the great deep, be considered as a remnant of the archaic story of the Deluge which referred to the water as coming from the eyes of the goddess Tiamat. After Tiamat was cut into two parts, the rains were still considered to be coming from her body but from the holes in her upper part, which formed the skies (Cassuto 1983).

The ancient tradition of Tiamat as a monster of the deep seas is found also in the Bible: ***Awake, awake, put on strength, O arm of the Lord awake, as in the ancient days, in the generations of old. Art thou not it that hath cut Rahab,***

Fig. 6.2 A pool formed by a spring on the foothills of the Zagros Mountains (Photo by Issar 1964)

Fig. 6.3 The Ararat Mountain, a view from the south (Photo Issar summer 1963)

[and] wounded the dragon?: Art thou not it which hath dried the sea, the waters of the great deep (tehom Raba) (Isaiah 51:9–10)

Another detail of the story of the Deluge is the landing of The Ark on the mountain of Ararat. This mountain, which is part of a chain of mountains, is located on the border between Turkey, Iran and Armenia, far away from the plains of Mesopotamia. Whether these are exactly the ones that the ancient narrator had in mind is not sure. Yet this mountain is very impressive when looked upon from the south (Fig. 6.3)

The higher one (Fig. 6.3) reaches an altitude of 5,205 meters. Its top is permanently covered by snow (photo shown in Fig. 6.3 was taken in August) and glaciers occur in some valleys. One can thus conclude that during colder periods the glaciers moved downwards, covering everything in their path, and also causing the deposition of typical glacial sediments. When they retreated, at the end of the glacial periods, the sediments were eroded, exposing buried trees, like the ones found in such sediments in the Alps. These were, most probably, interpreted by the ancients as remnants of the Ark.

As mentioned the story of Marduk defeating Tiamat and her family in *Enuma-Elish* was chanted during the New Year Festival in Babylon at the beginning of the springtime, with the appearance of the crescent-moon at the first of the month Nissan, the first month of the Babylonian Calendar. The holy-day began with the ceremony of the cancellation of the coronation of the king, and ended with the ceremony of his re-coronation. The New Year festivals were accompanied by ten days that were days of repentance and asking forgiveness in order to protect the human from the natural forces of destruction. They were celebrated at the beginning of the spring or at the end of the winter, which is the time of the beginning of the ripening of the grain and the greening of the pastures, the blooming of the fruit trees and the birth of lambs.

References

Bar-Matthews M, Ayalon A (2011) Mid-Holocene climate variations revealed by high-resolution spelepthem records from Soreq Cave, Israel and their correlation with cultural changes. Holocene 21(1):163–171

Baruch U (1986) The Late Holocene vegetational history of Lake Kinneret (Sea of Galilee) Israel, Paleorient 12(2):37–48

Cassuto U (1983) A commentary on the Book of Genesis. Magness Press, Hebrew University, Jerusalem

Degens ET, Wong HK, Kempe S et al (1984) A geological study of Lake Van, Eastern Turkey. Geologische Rundschau 73:701–734

Frumkin A, Magaritz M et al (1991) The Holocene climatic record of the salt caves of Mount Sedom, Israel. Holocene 1(3):191

Hallo WW, Simpson WK (1971) The ancient Near East, a history. Harcort Brace Jovanovich, New York

Hammer CU, Clausen HB et al (1980) Greenland ice sheet evidence of post-glacial volcanism and its climatic impact. Nature 288:230–235

Hammer CU, Clausen HB et al (1987) The Minoan eruption of Santorini in Greece dated to 1645 BC. Nature 328:517–519

Heidel A (1949) The Gilgamesh Epic and Old Testament Parallels. University Chicago Press, Chicago

Issar AS, Sorek S (2010) The possible dimension, additional to space time, which physicists ignore. J Mod Phys 1:70–76

Karlen W (1991) Glacier fluctuations in Scandinavia, in temperate palaeohydrology. In: Starkel L, Gregory KJ et al (eds) Fluvial processes in the temperate zone during the last 15,000 years. Wiley, New York

Lambert WG, Millard AR (1969) Atra-Hasis, the Babylonian story of the flood. (Also the Sumerian Flood Story by M. Civil) Clarendon, Oxford

Landmann G, Reimer A et al (1996) Climatically induced lake level changes at Lake Van, Turkey, during the Pleistocene/Holocene Transition. Global Biogeochem Cycles 10(4):797–808

Lemcke G, Sturm M (1997) d18O and trace element as proxy for the reconstruction of climate changes at Lake Van (Turkey): preliminary results. In: Dalfes H, Nüzhet et al (eds) Third millennium B.C. climatic change and old world collapse. Series: NATO ASI Series (closed), vol. 49, Springer-Verlag, Berlin

Pritchard JB (ed.) (1958) The ancient near east, an anthology of texts and pictures. Princeton University Press, Princeton

Reade J (1983) Assyrian Sculpture. British Museum Publications, London

Schoell M (1978) Oxygen isotope analysis on authigenic carbonates from Lake Van sediments and their possible bearing on the climate of the past 10,000 years. In: Degens ET, Kurtman F (eds) The geology of Lake Van. The Mineral Research and Exploration Institute of Turkey, Ankara

Smith G (1876) The chaldean account of genesis. Scribner, Armstrong and Co. New York

Stiller M, Ehrlich A et al (1984) The Late Holocene sediments of Lake Kinneret (Israel)-multidisciplinary study of a five meter core. Geological Survey of Israel. In: Geological Survey of Israel, Ministry of Energy and Infrastructure, Jerusalem

Woolley L (1929–1954) Ur of the Chaldees. Pelican, Gretna

Chapter 7
From the Tower of Babel to the Laws of Hammurabi

Abstract The story of the Tower of Babel is an explanation, by the ancients for the variety of languages, as well as the existence of the tall ziggurats, i.e. temple-towers, characterizing the landscape of the valley of the Tigris and Euphrates. The lower stretches of Mesopotamia are arid and only irrigated agriculture could be practiced. Thus the settlement of this region by the Sumerians started only after man learnt to divert water from the big rivers. This happened during the Late Neolithic to Early Chalcolithic termed the Hassuna-Sammra periods at ca. 7500 B.P., followed by the Ubaid period, which began at about 6900 B.P. and continued till about 5300 B.P. It was followed by the Uruk period, fourth millennia B.P. during which writing was invented. Between ca. 4 K and 3.9 K B.P. Mesopotamia went through a series of invasions. Hammurabi at about 3.8 K B.P., established the Babylonian-Akkadian civilization by amalgamating the different religious traditions and laws into a state religion and civil code. The prevailing theory blaming the people of Sumer for the economic crisis of the period around 4 K B.P. is refuted, as the cause was a severe reduction in the flow of the rivers due to desiccation of the climate.

> and it came to pass, as they journeyed from the east, that they found a plain in the land of Shinar; and they dwelt there. And they said to one another, go to, let us make bricks, and burn them thoroughly; and they had brick for stone, and slime had they for mortar. (Genesis 11: 2, 3).

After the water of the Deluge retreated and Noah and his family and all the "living substance", which survived left the Ark, the phase of revivification of life upon earth had to start. We are not told about the evolution of the family tree of the "living substance" a task left for Charles Darwin and his followers. The Bible focuses on the evolution of humankind, starting from Noah's three sons Shem, Ham and Japheth (and of course their nameless wives).

The array of the genealogical tree of mankind (Genesis, 10:1–32 & 11:10–32) is given to explain to the reader of the Bible the branch which produced the father of the Jewish people namely Abraham the son of Terah. The reader is informed that from Noah's son Shem sprang five secondary branches of Semites, which

A. S. Issar, *Strike the Rock and There Shall Come Water*, SpringerBriefs in Geography, DOI: 10.1007/978-3-319-01937-6_7, © The Author(s) 2014

gives an argument of self defense to all anti-Semites to claim that they are not just anti-Jewish but have a few more targets to their racial hatred. Moreover, to answer the question why is their hatred concentrated against the Jewish people their self-justification will be that all the names of the other descendant of Shem are not mentioned. Indeed, Shem's only heirs mentioned are Aram, who had four sons, and Arphaxad who was Eber's grandfather, from whom stem the Hebrews. For the time being, inspired by the biblical genealogical tree, the focus in this chapter will be put on the descendant of Shem, after six generations, namely Abraham the son of Terah.

Yet before the story of Abraham the son Terah is reached one can not skip the story of the Tower of Babel, which comes after the story of Noah's Deluge. In the opinion of the present author, this story is additional evidence of the tendency of the ancients to give a logical explanation to the questions stemming from their observations.

In the first place the writers of the Bible had to explain the variety of languages, characterizing the mosaic of nations inhabiting their *terra cognita*. The second observation was the tall ziggurats jutting from the plain of the valley of the two rivers. These ziggurats were temple-towers built by the Sumerians, Babylonians and Assyrians.

These towers built of bricks needed a logical explanation to the people, like the Children of Israel who had been expelled from their mountainous country. The most appropriate explanation would be that which answers two questions, namely that of the profusion of languages and that of the height of these towers. The optimal answer is the biblical story of the Tower of Babel. Reading this story with the appreciation of the observing eyes and enquiring minds of the ancients fills the reader with admiration:

We recommend re-reading this story, Genesis 11:1–10, with the present author's comments in brackets.

> And the whole earth was of one language, and of one speech.

Explanation by the present author: The story starts with the logical conclusion from the story of a global deluge, which is that just one family came out from the ark. From this "fact" arises the question: How is it that after a rather short time since the deluge there is such a multitude of languages? And here comes the story, which explains the phenomenon of many languages and the field observation of high towers.

> Genesis 11: 2 And it came to pass, as they journeyed from the east, that they found a plain in the land of Shinar; and they dwelt there.

Ditto: The land of Shinar is most probably Sumer and the Ziggurat of Ur is the most impressive.

> Genesis 11: 3 And they said one to another, Go to, let us make brick, and burn them thoroughly. And they had brick for stone, and slime had they for mortar.

Ditto: The consequence of the observation of the writers of the story that the ziggurats are built of bricks made of burnt clay, while their ancestors' temples were built of stone.

> Genesis 11: 4 And they said, Go to, let us build us a city and a tower, whose top may reach unto heaven; and let us make us a name, lest we be scattered abroad upon the face of the whole earth.

Ditto: Answering the question: What was the purpose of building the ziggurats? The height of these towers could be explained only as a result of the wish of its builders to reach the sky, which indeed they were (virtually speaking).

Genesis 11: 5 And the LORD came down to see the city and the tower, which the children of men built.

Ditto: The term 'children of men' bears a tinge of scorn.

Genesis 11: 6 And the LORD said, Behold, the people is one, and they have all one language; and this they begin to do: and now nothing will be restrained from them, which they have imagined to do.

Ditto: Expressing the hidden wish for unity, which gives strength, and anxiety from division and disintegration of the society, be it family, tribe or nation.

Genesis 11: 7 Go to, let us go down, and there confound their language, that they may not understand one another's speech.

Ditto: The subconscious understanding of the writers of the story of the vital role for the survival of unity of any society of a mutual language which enables understanding between people.

Genesis 11: 8, 9 So the LORD scattered them abroad from thence upon the face of all the earth: and they left off to build the city. Therefore is the name of it called Babel; because the LORD did there confound the language of all the earth: and from thence did the LORD scatter them abroad upon the face of all the earth.

Ditto: The ancients who composed this story were not too pedantic with the grammar of the Hebrew language as the root of confounding is *bll* (read balol) and thus the derivation of Babel from this root is rather questionable, grammatically speaking. On the other hand the source of the name Babel as already mentioned in the Akkadian language Bab means door or gate, while Ilim, means gods, which seems more logical. But then the story of the Tower of Babel would have left too many observations and wonderings without a comprehensive answer.

Returning to the genealogical tree of Shem (Genesis 11:12–26) one learns that his children were Elam, and Asshur, and Arphaxad, and Lud, and Aram. No city or other ancient site bearing the name of one of Shem's descendants was mentioned again in the Bible except the name of Elam. It was the name of a kingdom of warriors in the west of Iran on the frontier of Mesopotamia between the third millennium B.C. and the sixth century B.C. Among its kings is mentioned Chedorlaomer, whom we'll discuss later. Until today a small town exists which is named Elam, in the Zagros Mountains in the west of Iran. The author was also there in the context of his research of the groundwater of this area. Of the Assyrian capital cities only ruins remain, but the name Assyria remained etched in blood and destruction in the history of the Ancient East. It is most probable that the city of Lud, located in central Israel, was not named after Lud son of Shem and not even after the "Ludim", which in verse 13 Chap. 10 of Genesis, the Bible associates them with the descendants of Egypt. Yet, to Lud at least a memory exists, while to *Arphaxad*, the son of Shem the grand father of Eber, not even a memory remains. In contrast to this, Arphaxad's brother, our uncle Aram, left a big impression in the history of our people and area.

As mentioned above, the lower stretches of the Mesopotamian plains are arid, and thus only irrigated agriculture could be practiced there. Thus the settlement of this region by land-tillers started only after man learnt to build canals and divert water from the big rivers. This happened during the Late Neolithic-Early Chalcolithic (namely, the Hassuna-Sammra periods at ca. 7500 B.P.). During the Late Chalcolithic one finds a profusion of settlements, all representing a well-defined culture (called Ubaid) which began at about 6900 B.P. and continued till about 5300 B.P. The people who practiced this culture most probably came from the southeast where they developed methods of irrigation from streams and springs flowing in the valleys of the Iranian plateau and later at about the middle of the Chalcolithic period, they arrived at a stage which enabled them to cope with the technical and organizational problems of irrigation from big rivers on a large scale. As stone is rare in the Mesopotamian plain these people made full use of the clays. Their pottery and bricks and also their sickles, axes, and knives were of burnt clay. Their Mother Goddess was also made of clay. The temples were built one on top of the other.

The first people who began to develop their civilization (called The Ubaid culture) in the land which was to be called Sumer (Most probably biblical Shinar) may have belonged to the aborigines of the valleys of Iran before the penetration of the Indo-Aryans. They settled in the flood plains of Mesopotamia and began to practice irrigation from the rivers to supply their rather large communities. They might have learnt these methods also from their neighbors in the northern part of Mesopotamia or may have brought them from the valleys extending from the Iranian Plateau. Their eastern beliefs and cults came in contact with those of the Semites from the west and of the aborigines, who may have occupied the northern part of Mesopotamia since Neolithic times, as shown by some ancient Mesopotamian place names, which are neither Semitic nor Sumerian.

The Ubaid culture was replaced by the Uruk culture, named after the city of Uruk (the biblical Erech) where it was first unearthed. At Eridu another Sumerian town and one of the most sacred, as it was the residence of Enki the God of Waters, 17 successive layers containing ruins of temples were found. Plans of the temples are identical. In each one of the layers of ruins a thick layer of fish bones is found on the floor of the temple, apparently food brought to Enki This succession of layers continues without interruption into the historical period.

The invention of writing occurred during the fourth millennia B.P. (during the Uruk period). It developed into the cuneiform script which was used until the end of the Achaemenide period, namely the time of Alexander the Great of Macedonia (ca.300 B.C., 2300 B.P.).

The civilization which started in the southern part of Mesopotamia was organized in large communities in the beginning in city-states, later in kingdoms which dominated vast areas, and finally expanded into empires. As can be seen from the archaeological record, the colonization of northern Mesopotamia began in the Lower Neolithic period and reached southern Mesopotamia only during the Upper Neolithic-Early Chalcolithic, yet, in later periods the people of the south ruled the north. This was most probably due to the fact that the southerners had to organize bigger governmental systems to cope with the basic problems of their irrigation systems, while the northerners could make do with the natural precipitation supply. The building and maintenance of the irrigation systems brought with it a system of administration supported by a military to levy taxes, build control and protect the water projects.

Thus the Sumerians, who came most probably from the east occupying the southern part of Mesopotamia, and the Semites, who came from the west and north-west occupying the central part, shared the same big valley and were in close contact. They exchanged world views and traditions while developing their written literature. It seems quite obvious that they started from different archaic, most probably prehistoric, religions and cultures, as mentioned above, and as will be discussed further in more detail. But living together in the same valley and sharing the same environmental conditions caused them to develop a similar culture. As was demonstrated in the chapters dealing with the myths of creation and the Deluge, the Akkadians adopted quite a lot of the earlier Sumerian myths, but they also brought some of their own with them, which they changed to better fit the framework of the new environment. It seems that the analogy found between the myths in the Akkadian and biblical texts is due, in the first place, to a common pre-Mesopotamian heritage, but also to a blending process which took place especially after Sumer was conquered by the Semites, as will be told later.

The Sumerians were the first to build large water projects connected with their ability to build the first big organized settlements. In about 5000 B.P. Uruk had a population of about 50,000 inhabitants while Old Uruk in about 4800 B.P. had about 34,000 inhabitants (Kramer 1954). The pivot of the organization was the local god, its temple, and the religious ceremonies which took place according to the changing seasons. The king and ruler of the city was a vassal of the divine power. (Fig 7.1)

One of the most important aspects of the ancient world which developed in Sumer in connection with the big irrigation projects was slave labor. The larger the public projects became, the more man-power was needed for their execution and maintenance. The hierarchical pyramid became more and more established as base layers became more numerous, its top richer, the in-between layers more dependent on the wealth of those at the top, and the work and subordination of those at the bottom. This brought about the evolution of hierarchical and despotic governmental systems characteristic of the big river civilizations.

A rather short cultural transition time (named the Proto-literate period) marked the last steps of emergence from prehistory to history. As the invention of writing spread, the Uruk culture was replaced by the Sumerian Early Dynastic kingdoms, the history of which was written on the cuneiform tablets found in the excavation in the Mesopotamian Valley.

The high-resolution interpretation of the Soreq Cave, shows that the collapse of the Uruk society is a result of a period of dryness (Bar-Matthews, Ayalon 2011).

As mentioned during the discussion of the date of the Deluge, it may well be that during the Proto-literate or even the during Uruk times (around 5000 B.P.) a more humid climatic change took place which produced a series of floods destroying towns and fields and which was afterwards explained as the global catastrophe of the Deluge.

Fig. 7.1 Map of ancient Near East (Based on map produced by 'Bible History Online')

The historical records found in Sumer start the list of kings from the Deluge on. This list of kings also includes Gilgamesh, the King of Uruk, whose search for immortality brought him to hear the story of the Sumerian, Ut-Napishtim the survivor of the Deluge. Thus, at about 4750 B.P. the period began of the Sumerian city-states ruled by the post Deluge kings or as it is also termed, the Early Dynastic Period.

The Land of Sumer was an agglomerate of city-kingdoms, strung along the Euphrates from the latitude of Babylon to the Persian Gulf (Fig. 7.2).

Each city-kingdom was surrounded by satellite towns and villages situated between fields (*gan* in Sumerian), orchards and palm groves irrigated by canals from the river. The open uncultivated steppe between the irrigated areas served as

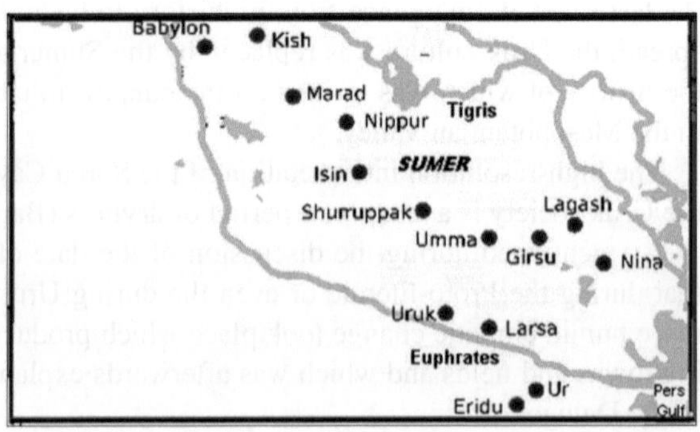

Fig. 7.2 Map of Sumer and location of its cities

pasture land and was called in Sumerian *edin*. Thus natural non-cultivated stretches of land were called *Gan-Edin* in Sumerian. This reminds one of the biblical *Gan-Eden* (Garden of Eden), the archaic garden where man lived so happily before he ate the apple offered by his wife and had to toil on his land for his bread. The pivot of each city-kingdom was its god-protector. Enlil, the chief Sumerian god, was the patron of Nippur. Uruk was dedicated to Ann (Anu), the sky god, and Inanna (Semitic Ishtar), the Goddess of Love. Ur belonged to Nanna (Sin of the Semites) the Moon God and Lagash to Ninqirsu son of Enlil. The god of Larsa was the sun god Utu (Semitic Shamash). Enki whose domain was complex, was the protector of Eridu. Literally, his name means "lord earth", but possibly because the earth without water had no use, Enki also became the deity of the fresh water. At the same time he controlled the flow in rivers and canals, the rise in springs and wells. The cultivated land around the cities were divided into three parts: one part, the land of the lord (*gana-ni-en-na*), was worked by all the community for the sole benefit of the temple; another part, the food land (*gana-ku-ra*), was allotted to the dependents of the temple; the third part, the ploughed land (*gana-apin-la*), was let to tenants against one seventh or one eighth of the harvest. Thus, the temples had at their disposal high revenues which were used also to finance large-scale operations, building of temples, fortifications, and irrigation canals. The ruler of the city-kingdom was thought to be chosen by the city-god and was, in the archaic period, also the chief priest. He was also considered to be the personification of the god. Afterwards the tasks were divided between king and priest, but the king (Ensi or Lu-gal) was the one responsible before the god for the well-being of the people of the city.

War between the cities was for possession of the irrigated lands. Success and victory meant many irrigated lands, revenues and power. Thus after a period characterized by the rule of city kings, groups of more than a few cities were ruled by one victorious king. The Ensi of Umma at the command of his god Enlil raided the irrigated land, the fields beloved of Nir-gir-su. He ripped out the stele which marked the border between fields, cities, and kingdoms, and entered the plain of Lagash. The King of Umma went on to build a state kingdom, he conquered all Sumer and claimed even to reach beyond its border: "When Enlil king of all sovereign countries had given him the kingship over the nation… Enlil let him have no opponent, all sovereign countries lay (as cows) in pasture under him; the nation was watering its field in joy under him" (Roux 1966 ditto).

This Sumerian kingdom did not last for long. Sargon the Akkadian King of Kish, conquered Sumer in 2371 B.C. and established an empire from Iran to the Mediterranean, ruled by the Semitic Akkadians. This empire which lasted, although in a reduced area, until 2230 B.C., caused the expansion of the Semitic language and cuneiform script over most of the Middle East. Trade and exchange of merchandise, as well as technologies, extended from the Hindus Valley and plain (where another wonderful irrigation civilization emerged) to Anatolia, Greece and Egypt.

The second Sumerian period started about 2100 B.C. after an interval of about 100 years during which the Mesopotamian valley was dominated by the Gutians, warring tribes coming from the Zagros Mountains. During this Neo-Sumerian reemergence, the city of Ur under King Ur-Nammu seized hegemony over the other cities. This empire later expanded beyond the borders of the valley, governing a vast territory by local En-sis, Sumerians or Akkadians ruling as vassals to the great king. This period, which was peaceful inside the empire, although war persisted along its northeastern borders, was the period of the building of the gigantic ziggurats an echo of which one finds in the biblical story of the Tower of Babel.

In the year 2006 B.C Ur the capital of the Sumerian kingdom, was destroyed by the Elamite tribes living along the Zagros foothills. This destruction was the end of Sumer as an authentic independent society and culture. The stage was cleared for another society of western Semites known as Amorites, or in Sumerian *Mar-to* or *Mura-tu* namely westerners. These people came from the desert, western part of the Fertile Crescent, penetrated into Mesopotamia, conquered many cities and settled down in and around them. They adopted the parts of the Sumerian beliefs and way of life which were to their liking and which they could make use of. Although the *coup de grace* to the capital Ur was given by the Elamites, it was only a long process of deterioration of the kingdom and the continuous encroachment of the Mura-tu, which finally brought about its collapse. The history of Sumer became buried under the ruins of its cities, turning into clays due to the disintegration of the mud bricks.

The downfall of the Sumerian kingdom as a result of the invasion of the people from the west is well documented by clay tablets found in the archives of Mari located on the western bank of the Euphrates and in the excavated ruins of Ur. These clay tablets portray the sequence of events which has recurred many times in history of the collapse of a well-established, sedentary, civilized society under the attack of "non-civilized" nomadic and semi-nomadic tribes. The latter being dependent on the random supply of rains, once this supply fails, their only possibility to survive is to encroach on the fertile plains irrigated by the water of the rather stable big rivers where food, pasture, and other trophies were abundant. As already mentioned, nomadism, usually interconnected with the sedentary communities near springs, started to evolve as a way of life after the goat and sheep became domesticated, probably during the Neolithic period. The need to cover large areas in order to supply food for the livestock brought about the formation of a new type of society which existed side by side with that of the societies living on irrigated agriculture. The relations between the two depended on the climate. When this was favorable these two societies complemented each other from the economic point of view (and sometimes even the social by providing mercenaries) but once a period of droughts encroached over the desert, the nomadic society had no other resource but the fertile fields of the sown lands. When they could not obtain this land in cash or kind, they took it by force.

This clash between the pastoral nomad and the farmer is a story which repeats itself throughout the history of the Fertile Crescent. It is first heard in the biblical story of the conflict between Cain and Abel. It might have started with the drying up of the climate in the later part of the Neolithic around 6500 B.P. This caused, most probably, the penetration of agricultural societies into the more southern parts of the fertile plains of Mesopotamia, irrigable by water from the two rivers, and the lands along the rivers were practically empty. There is an indication of another aridization phase also at about 5000 B.P. (Mid–Early-Bronze Age) and another one which brought the invasion of the Amorites at about 4000 B.P.(End Early to mid Bronze Age).

In the curves presented in Figs. 1.2 and 1.5 one can see that at about 4000 B.P. the annual amount of rainfall in the vicinity of Jerusalem decreases, as deducted from the ratios of oxygen and carbon environmental isotopes of the stalagmites of the Soreq Cave (Bar-Matthews, et al. 1996). There is also a decline in the level of the Dead Sea (Bookman et al. 2004) At the same time there is a decrease in the humidity in the region of Lake Van as detected by Lemcke and Sturm (1997) and a retreat of the glaciers of Scandinavia (Karlen 1991).

In the core samples taken from the bottom of the Sea of Galilee there is a peak of reduction in the pollen of olives and an increase in that of oaks at 4000 B.P. (Baruch 1986). These changes apparently show that, as the semi-arid environment of the Sea of Galilee became drier, man abandoned the olive orchards and the natural forest re-established itself. At present the iso-precipitation line of the Sea of Galilee is about 400 mm/year (Fig. 5.1) and, as the olive tree needs minimum 300 mm/year to produce fruits, it means that the amount of precipitation was less than that.

An additional observation telling about the warm climate phase around 4 K B.P. is that of Dr. Avner Raban (1987) on settlements along the Israeli coast during the Bronze Age. In addition to his archaeological findings, Dr. Raban reports that around 2000 B.C. i.e. ca.4 K B.P., occurred a rise in the sea level and a rise in the supply of silt and sand which caused the silting up of the river outlets. This forced the people who used the rivers' outlets as ports, to excavate artificial outlets. The rise in the sea level is due to the melting of the glaciers (see Karlen's curve Fig. 1.5) in the cold regions, while the rise in the supply of silt and sand is a result of high floods of the Nile caused by increased precipitation rates over its drainage basin in the mid-latitude zones due the warming of the oceans.

This regional aridization caused again the nomadic and semi-nomadic societies to escape famine by invading the green and fertile valleys. This time, however, unlike their earlier encroachments, they met well-established societies, the people unwilling to share their food and possessions with the newcomers. Thus a series of clashes and wars started, causing the downfall of the Sumerian kingdom and the seizure of power by the invaders. Once the invaders became the rulers, they divided the lands, temples, and treasures of the deposed former ruling class among themselves and slowly assimilated with the sedentary population, which continued to cultivate the land, an occupation the former nomads were not willing to adopt.

The conquering people who came from the west were Semites and as such had a heritage different from the Sumerians but similar in many aspects to that of the Akkadians. The Semitic culture the conquering Amorites brought with them, evolved east of Mesopotamia during the Chalcolithic at the end of the Early Bronze Age (from about 5.5 K B.P., to about 4 K B.P). As already stated, they succeeded in establishing themselves in southern Mesopotamia seizing power and control over the sedentary societies. Thus a new ruling class of kings and

En-sis of non-Sumerian origin established itself in Sumer and soon adopted the Sumerian Akkadian culture. As their language was similar to the Akkadian they went through a process of assimilation which led to the formation of a Semitic-Sumerian culture that succeeded in replacing the original Sumerian although it adopted its script and many of its ancient customs. Due to a process of bi-lateral assimilation Sumerian deities and myths were adopted and blended into Semitic mythology and rites. The tasks and dominion of the gods were changed here and there to agree better with the original Semitic gods, whose names were maintained. These gods, after whose names the rulers were called, presented another agricultural system dependent more on rain than on flood, thus the name of the god Adad, who was the controller of the rain, became prominent beside El, the chief god and his consort Ishtar-Goddess of sex and fertility and Dagan-God of earth fertility.

A series of waves of repeated invasions of Amorites to Mesopotamia continued, one after another. The newcomers, driven out of their country by the continuing dryness and thus famine, tried to gain their share in the spoils of the rich lands. However, after replacing the ruling class and becoming established economically, the first wave found more interests in common with the original society over which it was ruling than with the new arrivals. After two or three generations the Amorite, the Sumerian and the Akkadian found themselves in need of fighting back the new waves of invasions by people from the west. Thus, between about 4 K and 3.9 K B.P. Mesopotamia went through a series of invasions and establishment of small kingdoms ruled by the descendants of the newcomers who fought back the new invasions of Amorites. After establishing themselves in the lands they conquered, the new rulers started to fight each other. Each tried to encroach on the lands of the other until one of the new dynasties succeeded in overpowering the others and established a new empire.

Hammurabi was the leader, who brought this process to its climax, at about 3.8 K B.P. (1780 B.C.). He established a Babylonian empire and the Babylonian-Akkadian civilization. This was done by amalgamating the different religious traditions and laws into a state religion and civil code. Marduk was proclaimed the chief god and was accepted by the former gods as their savior from the forces of chaos and destruction. The story of creation, the Enuma Elish, already mentioned, as well as the other myths and traditions, were adopted and codified. The symbols of order were, of course, the laws, religious as well as civil. The famous code of Hammurabi contains many laws which are similar to laws found in the Bible. The detailed comparison between the two systems of laws is beyond the scope of the present study, which concentrates on the hydrological aspects of history, yet it will be shown later, that though the Hebrew law was in some ways influenced by the Babylonian one, the two systems represent two altogether different socio-economical environments. At the same time, one gains from the Laws of Hammurabi interesting glimpses of the life of the people in Mesopotamia which are important in order to understand the civilization which had still much in common with that of the Hebrews.

The interesting aspect from the hydraulic point of view is that while the laws dealing with criminal behavior of the citizens have parallels in the code of Moses, those which deal with land do not. Thus one can assume that while the criminal code was part of the Amorite-Semitic heritage the land code was adopted from the ancient Sumerian law.

Moreover, the criminal laws (Pritchard 1958, McNeil et al. 1968), which are most similar to those found in the Bible are feasible in a nomadic society as well as a sedentary one, while the land and water laws are, of course, suited to a farming community only. Also some of the criminal laws show the influence of a river society. Thus, while the first law is the following: (1)"If an *awelum* (citizen) accused another *awelum* and brought a charge of murder against him but has not proven it, his accuser shall be put to death". Such a law can be compared to that of Moses' code "But if a man come presumptuously upon his neighbor, to slay him with guile, thou shalt take him from mine altar, that he may die" (Exodus 21:14); on the other hand, the second law of Hammurabi states the following: (2) "If an *awelum* brought a charge of sorcery against another *awelum* but has not proven it, the one against whom the charge of sorcery was brought, upon going to the river, shall throw himself into the river and if the river has then overpowered him, his accuser shall take over his estate, if the river shows the *awelum* to be innocent and he has accordingly come forth safe, the one who brought the charge of sorcery against him shall be put to death, while the one who threw himself into the river shall take over the estate of his accuser."

It is quite obvious that two different world views, stemming from two different environments, can be seen here, the first of which may be the law of any society. Its resemblance with that of the Hebrews may point to a Semitic origin. The second law is that of a society where "the river", in this case, the Euphrates, is regarded as a divine entity which decides the fate of man.

Other examples of the sort of laws found in that of Hammurabi but not in the Bible are the following:

(53) If an *awelum* was too lazy to make the dike of his field strong and did not make his dike strong and a break has opened up in his dike and he has accordingly let the water ravage the farmland the *awelum* in whose dike the break was opened shall make good the grain that he allowed to be destroyed.

(54) If he is not able to make good the grain, they shall sell him and his goods, and the farmers whose grain the water carried off shall divide.

(55) If an *awelum* upon opening his canal for irrigation became so lazy that he has let the water ravage a field adjoining his, he shall measure out grain on the basis of those adjoining his.

(56) If an *awelum* opened up the water and then has let the water carry off the work done on a field adjoining his, he shall measure out ten *kur* of grain per eighteen *tha*.

These were the most obvious cases which show the Mesopotamian sources of the Laws of Hammurabi. Also the other laws which have nothing to do with irrigation and water show quite clearly that the suggestion of seeing the code of Hammurabi as a precursor of the code of Moses does not take into consideration the basic differences in environment, society, and spirit, not to speak of the religious aspects of the two codes of laws.

The personal involvement of Hammurabi in the technical problems of water supply can be learnt from a few letters written on clay tablets in which he gives detailed instructions to dredge and clean certain canals, in order to secure the income from the rented fields. In one of the letters he instructs an official by the name of Shamash-Azir (helper of the Sun God) to investigate a complaint of a tenant that the level of a certain field is above that of the canal and cannot be irrigated. In case the complaint is true, he should be given another irrigable parcel belonging to the fields of the Palace, and "let him have no reason to withhold the barley which he pays for rent" (Newman 1980).

The death of Hammurabi (1750 B.C.) was also the beginning of the end of his Babylonian empire, caused by internal as well as external forces. The people of the former kingdoms of Mesopotamia revolted and in addition there followed invasions by peoples from the east and northwest. The invaders this time were of Indo-Aryan origin, namely Kassites and Hittites.

In the year 1595 B.C., Babylon was conquered by the Hittite king Mursili I. The King, Samsu-ditana, was killed and the statue of Marduk was taken as booty. After the Hittites left Babylon, most probably because of internal palace intrigues against the campaigning king, Babylon came under the rule of the Kassites, who later promoted another Babylonian-Kassite empire to which we will return later.

Another archive excavated by the archaeologists is that of Larsa. In the following paragraphs the archive of public waterworks will be discussed.

The kingdom of Larsa was established by an Amorite chieftain by the name of Nablanum in the year 2025 B.C. After four generations his descendant Gungunum, in the year 1924 B.C., overpowered Lipit Ishtar the ruler of the adjacent kingdom of Isin, which had been established by another Amorite chieftain. The annexation of Isin enabled the establishment of a new empire which extended over the area which comprised most of ancient Sumer. During the reign of Gungunum's son, King Abisare (1905–1895 B.C.) and grandson, King Sumuel (1894–1860 B.C.), new irrigation projects were built and others were maintained. A description of these works is found in an archive of official records which consists of letters and documents written on clay tablets in the Akkadian language with many Sumerian terms. The tablets mention mainly a certain official by the name of Lu-igisa who is entrusted with certain public water projects and has to administer the works, payment, protection, and even surveying of new canals. The name of this official is Sumerian (meaning "a man with a kind face" or "the friendly benevolent one") but many of the other names, especially his superiors (not to speak of the king Sumuel) have Semitic names such as Nur-Samas (the sun gods of the Western Semite) or a lady called Arnabum ("Bunny" or Arnevet in Hebrew). From the translated tablets (Walters 1970) it seems that although the governing class was Semitic, the administration, irrigation and farming traditions remained Sumerian.

At Larsa, the Irrigation Bureau was headed by Nur-Sin, who was responsible to King Sumuel for all matters regarding the supply of water for irrigation. Under him served Lu-igisa, who seemed to have been a man for all purposes. A few of these records are presented here for the reader to share the echoes of ancient times.

> To Lu-igisa speak. This is what Nur-Sin says: The side canal which I and you surveyed, the son of Ku-Nanna is about to excavate. Before he begins, you excavate it and report to Isar-kubi (Walters 1970, p. 33).
> To Lu-igisa speak. This is what Nur-Sin says: Isar-kubi has just written me about emergency work needed at the Nubitar Canal. Here is his request: Hire 1800 workers to take care of you. Ask around, and pay ten "mana" of silver in order to hire workers. And until you can get away appoint some one of your choosing to supervise the crew. (Walters 1970, p. 35).

Other people with familiar names are mentioned in other tablets for instance:

> To Banum speak. This is what Abaraam says: You need to know that E-Kibi and Nabi-Enlil have just sent 5 shekels of gold intended for the canal to Isar-kubi in order to eat your grits (take advantage of you). Take action or Isar-Kubi will confront you. (Walters 1970, p. 109).

From the other tablets a rather clear picture emerges of how dependent the people of the kingdom of Larsa and, as a matter of fact, all the peoples of southern arid Mesopotamia were on the irrigation network. These canals were the main arteries which supplied the water for irrigation. They had to be dug and continuously cleaned so that they would not become blocked by silt brought by the rivers. This was the task of the administration. At the top of the hierarchical ladder was the King, Sumuel. It was he who decided about the big projects as well as allocation of the funds needed for it. He was also the arbiter and judge to whom the heads of projects had to report or submit their claims: "If you think I have responded falsely I will even respond to Sumuel my lord." (Walters 1970, p. 107). The chain of hierarchy under the king started from the head of the Irrigation Bureau Nur-Sin, who was most probably an Amorite. Under him served inspectors who had Amorite as well as Sumerian names like Lu-igisa, labor contractors, suppliers, foremen, workers and soldiers. The work of the Bureau for Waterworks consisted not only of digging and cleaning the canals but also preparing some fields for irrigation by furrowing them, and seeing that money allocated for work contracted should not be misappropriated.

The names of the kings and people of Larsa are quite familiar to those who know the Bible. An even more familiar connotation can be found in the Laws of Hammurabi. As mentioned above, the waves of Semitic-Amorite tribes encroaching over Mesopotamia continued also after the people of the first wave succeeded in establishing themselves as ruling dynasties over the people of Sumer. In the first year of Sumuel King of Larsa (1894 B.C.), another Amorite chieftain by the name of Sumu-Abum established a small kingdom in the Sumerian town of Ka-Dingir-Ra which was translated into Akkadian as Bab-ilim, namely, the gate of gods, or Bab-El, the gate of god. (The reader is reminded that in the Bible the reason behind this name is different and is connected with the language of the builders of the Tower of Babylon being confounded (Genesis 11: 9). The descendants of Sumu-Abum, slowly widened the domain of their country until in 1793 B.C. Hammurabi (Hammu—a Semitic deity, Rabbi—great) ascended the throne. In a campaign which lasted 38 years, he successfully annexed all the kingdoms from Assur and Mari in the north to Larsa in the south.

Thus, the archive of Larsa as well as other documents, such as the Laws of Hammurabi, portray all the features of a sedentary, established society based on river irrigation practices. This only about hundred years after the invaders took control. It can be concluded that the Amorite people, although a minority, were clever enough not to destroy the old system but to get the best out of it by adopting its administrative organization and methods, though changing its language as well as adding new ideas, gods, and traditions to the existing cultural world. After they settled down the Amorites too, became a society dependant on the river as the source of water for irrigation.

Before leaving Mesopotamia and migrating with the author's fore–fore...... fathers to Canaan and from there to Egypt, it is worthwhile discussing in more detail, the prevailing theory blaming the people of Sumer for the economic crisis of the period around 4 K B.P. (2000 B.C.). This theory claimed that the people of Sumer over-irrigated their land, leading to its becoming saline and as a result the decline in production and thus to an economic catastrophe. This suggestion by the archaeologist Jacobsen (1958, 1960) was based on three main sources of evidence:

1. Records from clay tablets from this period on soils becoming saline.
2. The shift from wheat to barley. Whereas in 2500 B.C. wheat was grown in equal amounts to barley and as late as 2400 B.C. it was still being raised in the proportion of 1: 6 to barley, wheat accounted for less than 2 % of the grain crop in 2100 B.C. and is missing altogether after 2000 B.C. As barley is more salt-tolerant, Jacobsen argues that salt encroachment caused this shift.
3. Reduction in crop yields. Estimates based on the records found in the Sumerian and Akkadian tablets, show that in 2400 B.C. the yield was 2,537 liters per hectare, in 2100 B.C. the yield was 1,460 liters per hectare and in 1700 B.C. only 897 liters per hectare.

Accordingly Jacobsen suggests that the change from wheat to barley and the reduction in the yields was the result of the soils becoming saline. This happened, in his opinion, due to the addition of large quantities of irrigation water to the area, which caused the water table to rise and come near the land surface, then by the sucking upward through the fine pores of the clayey soil reached the surface, evaporated, and left its salt behind.

Although, in principal, this anthropogenic process may have happened, it is still questionable whether it caused the reduction in yield or whether there were not some other natural processes that caused the agricultural system of Mesopotamia to decline. The biggest decline in yield was from 2100 B.C. to 1700 B.C. which, as mentioned already, was also the time of a dry period which affected the entire Middle East and may have caused a reduction in the total yield of the rivers and thus in the availability of water for irrigation. Moreover salinization due to a regional rise in the water table is connected with quite clear environmental phenomena on a regional scale, yet there is no mention of these phenomena in the Larsa water archives. If this process had been as crucial as suggested, one would expect to find it being mentioned more frequently and specifically in these archives which, as was shown, specify problems and methods in much detail. The alternative reason suggested by the present author, namely a severe reduction in the flow of the rivers due to desiccation of the climate, which brought a decrease in the amounts of water available for irrigation in addition to an increase in demand as a result of increase in population, may have caused the same environmental results. These will be local patches of soil becoming saline, as well as a shift from wheat to barley, which is drought-resistant as well as salt-resistant. The reduction in yields may also be due to smaller amounts of water applied in irrigation. It is thus suggested that, until more evidence is found in ancient documents, the blame put on the Mesopotamian people for their misuse of the water and the spoiling of their land should be taken only as a hypothesis rather than as a proven fact.

References

Bar-Matthews M, Ayalon A et al (1997) Late quaternary paleoclimate in the eastern mediterranean region from stable isotope analysis of Speleothems at Soreq Cave, Israel. Quatern Res 47(2):155–168

Bar-Matthews M, Ayalon A (2011) Mid-Holocene climate variations revealed by high resolution spelepthem records from Soreq Cave, Israel and their correlation with cultural changes. Holocene 21(1):163–171

Baruch U (1986) The Late Holocene vegetational history of Lake Kinneret (Sea of Galilee) Israel, Paleorient 12(2):37–48

Bookman (Ken-Tor) R, Enzel Y et al (2004) Late Holocene lake levels of the Dead Sea. Geol Soc Amer Bull 116(5/6):555–571

Jacobsen T (1958) Salt and silt in ancient Mesopotamian agriculture. Science 128:1251–1258

Jacobsen T (1960) The waters of Ur Iraq. Science 22:174–185

Karlen W (1991) Glacier fluctuations in Scandinavia in temperate palaeohydrology in fluvial processes. In: Starkel L, Gregory KJ, Thornes JB (eds) The temperate zone during the last 15000 years. Wiley, New York

Kramer SN (1954) History begins at Sumer. Doubleday Anchor Books, NY

Lemcke G, Sturm M (1997) d18O and trace element as proxy for the reconstruction of climate changes at Lake Van (Turkey): preliminary results. In: Third millennium B.C. climatic change and old world collapse. In: Dalfes H, Nüzhet et al (eds) Series: NATO ASI series (closed), vol 49. Springer Verlag, Berlin

McNeil WH, Sedlar JW (eds) (1968) The Ancient Near East. Oxford University Press, Oxford, p 261

Neuman J (1980) Five letters from and to Hammurapi, King of Babylon (1792–1750 B.C.) on water works and irrigation. J Hydrol 47:393–397

Pritchard JB (ed) (1958) The ancient near east, an anthology of texts and pictures. Princeton University Press, Princeton

Raban A (1987) Alternated river courses during the Bronze Age along the Israeli coastline. Colloque Internationeaux C.N.R.S. Deplacements des lignes de rivages en Mediterranee. Edition du C.N.R.S. Paris, 173–189

Roux G (1966) Ancient Iraq. Pelican Books, Gretna

Walters SD (1970) Water for Larsa, an Old Babylonian archive dealing with irrigation. Yale University Press, New Haven

Chapter 8
From Shem the Son of Noah to Abraham Son of Terah

Abstract Ur-Kassdim the city from which Terah and later of his son Abram-Abraham immigrated to Canaan is identified with Urkesh the capital of the Hurrian kingdom in northeastern Syria. The reason for this immigration was a period of dryness, which had affected the Middle East, starting around 2300 B.C. and reaching its peak around 1800 B.C. The wave of the immigrants also conquered a few of the Canaanite city-kingdoms and turned them to fortified cities. This climatic crisis led to the invasion of Egypt by the Hyksos, who gained control over northern Egypt. The immigration to Egypt was because of its prosperity due to the fact that the sources of the Nile are in tropical and sub-tropical Africa.

> Terah took Abraham his son, and Lot the son of Haran his son's son, and Sarai his daughter-in-law, his son Abram's wife, and they went forth with them from Ur of the Chaldees, to go into the Land of Canaan and they came into Haran and dwelt there (Genesis 11: 31).

The former chapter closed with challenging the accusation brought by the archaeologist Jacobsen of the people of Sumer for over irrigating their fields and thus causing the soils to become saline. The present chapter will open with challenging the identification by the archaeologist Sir Leonard Woolley of Sumerian Ur with Ur of the Chaldees from which ***Terah took Abraham his son etc.......from Ur of the Chaldees, to go into the Land of Canaan etc.*** (Genesis 11: 31). Woolley's identification, which was accepted for many years and quoted in many books, including the previous edition of the present book under the title *Water Shall Flow from the Rock* by the present author, should have been questioned in the first place, due to its location in the southern part of Mesopotamia, while the Bible tells us that Terah was on his way to Canaan. After the site of Haran was identified by archaeologists in Upper Mesopotamia near the modern village of Altınbaşak, in Turkey, the excavations at Tel Mozan in northeastern Syria, between the Euphrares and Tigris revealed an important late third millennium urban center found to be Urkish or Urkesh the capital of the Hurrian kingdom or Ur of the Hurrians (Buccellati, Kelly-Buccellati 1997, Berlyn 2005) Thus the travel from this city to Haran in order to reach Canaan seems more reasonable than that from Sumerian Ur. (Fig. 7.1).

A. S. Issar, *Strike the Rock and There Shall Come Water*, SpringerBriefs in Geography,
DOI: 10.1007/978-3-319-01937-6_8, © The Author(s) 2014

As a matter of fact, the blame of identifying Sumerian Ur with biblical Ur-Kassdim, from where Terah started his journey to Canaan, should not be put on the shoulders of the archaeologist Sir Leonard Woolley but on the scribes who wrote down the stories passed verbally from father to son on the history of the Hebrews. Indeed Sumerian Ur was inhabited during the rule of the Kassites over Mesopotamia. These people invaded Babylonia coming down from the Zagros Mountains after the fall of the Old Babylonian Empire after ca. 1531 B.C. and ruled it for about 400 years to about to ca. 1155 B.C. They also conquered the southern part of Mesopotamia, namely ancient Sumer down to Persian Gulf. The Kassites rebuilt the Sumerian city of Nippur and its ancient temples and it became the central city of the province. Other important centers under the Kassite rule were the Sumerian cities of Larsa, Sippar and Susa.

As mentioned the city which Terah left was Urkesh, which was abandoned at 1530 B.C. and forgotten until the archaeologist re discovered it, rather recently. When the scribe wrote down the stories this city was forgotten but Ur of the Khaldees-Kassites was still known, a fact which caused the scribe to force Terah and all his family to wander about thousand kms north-eastwards instead of traveling just a 150 kms from Urkesh westwards, to Haran.The excavations at Haran revealed that it was a major urban center during the third millennium B.C. and was the chief home of the Moon God Sin (Green 1992). Was this in the background of the ancient Hebrews who adopted the seventh day of the week, the quarter of the lunar month, as a holy day of rest and prayers?

One can learn about the inhabitants of the area south of the Taurus Mountains and west of the Zagros Mountains from the documents that were found in the archives of the city of Mari-Tel Hariri—on the western bank of the Middle Euphrates, today in Syria, about 300 kms south-east of Haran. (Fig. 7.1) Mari was a great and central city in an area inhabited by Semite tribes of Amoro-Amorites. Excavations were carried out by A. Parrot (1958, 1968) of a temple of the goddess Ishtar and a palace which contained a rich archive of clay tablets including correspondence with kings of Sumer when one of the kings of Mari became overlord over Sumer (Lloyd 1978). A cylinder seal inscribed with "Mesannipadda, King of Ur," which was probably presented to one of the kings of Mari confirmed the date, i.e. 2650–2400 B.C. These findings show that there were close connection between the northern city-states of Mesopotamia and the Land of Sumer.

From the archives of Mari it can be learnt also that the area east of Mari was occupied by the Amorite tribes of Hanu and Mar-Yamina or *Mar-Yamana*, in Akkadian as well as in Canaanite and Hebrew languages means "Sons of the South"., which may point to the fact that these were nomad tribes who migrated from the Desert of Syria, when the warm climate caused pasture and water resources to dry up. From the same archives it can also be learnt that around 1800 B.C. the Amorite chieftain Yahdun-Lim took over Mari, made it his capital and began a series of campaigns against the surrounding kingdoms and the chieftains of the Hanu and Mar-Yamina tribes. In the foundation inscription of the temple he erected to the sun god Shamash, he boasts about a campaign which had brought him to the shores of the Mediterranean and about the construction of irrigation channels in order to divert water from the Euphrates into the desert.

Returning to the people who left this region to go southward the question is: What were the reasons for this migration? The explanation suggested in this work is that the main reason was aridization, the same reason which caused the Amorites to invade Mesopotamia and later the Hyxos to invade Egypt. As mentioned above a climatic crisis of dryness had affected the Middle East, starting around 2300 B.C. reaching its peak around 1800 B.C. The severity of this crisis can be seen in the curve depicting the rainfall according to the stalagmites of the Soreq Cave the level of the Dead Sea, the humidity in the sediments of Lake Van as well as a retreat of the glaciers in Scandinavia (Figs. 1.2 and 1.5).

The influence of this crisis was noticeable especially along the borders of the desert, like the north of Mesopotamia. The average annual amount of the precipitations there was around 300 mm. From the geographical point of view these areas were near to the aridity line of 200 mm/year and their economy was based on irrigated agriculture and on grazing (Weiss et al. 1993). A decrease of a few percentages in the amount of the average annual rainfall, either as per year and/or higher frequency of drought years, brought an economic crisis. This caused the desertion of cities and to the migration of the population to the more humid areas or to the vicinity of the perennial rivers. The population of the semi-nomadic shepherds-farmers who lived in the rural environment surrounding the cities, benefited or suffered according to the state of affairs of the city. In the case of desiccation of the climate during, which started towards the end of the third millennium B.C. the climate crisis affected the urban as well as the rural population. The excavations at Subir (Tel Leilan), a city in north Mesopotamia (North east Syria), show that this city was abandoned around 1700 B.C., due to a marked increase in aridity. The echoes in the Bible of this climate change are the migration stories of the ancestors of the Hebrews. The area where Urkesh and Haran are located is a high plane that extends today in the south of Turkey, near the northern border of Syria, and is populated today by Kurds. The average annual rainfall in this area today is about 300 mm. Today, as in the past, agriculture in this area is possible only with irrigation. But as the Euphrates and the Tigris cross this region in deep channels, lifting the water to the level of the fields demands the building of dams or very long canals. A reduction in the rate of flow of these rivers, not to speak of the drying up of small springs and streams, will directly affect the population who is dependent on this water and the shepherds who benefit from the stubble. This population will be forced to migrate to areas where the regional desiccation had no critical consequences, either to the higher plains or to regions where tales of travelers say there is an abundance of water. In this case the global warm climate caused an increase in the precipitation over the subtropical regions and consequentially the level of the Nile. The nearest land to benefit is Egypt and rumors about the prosperity of this country may have reached the people on the northern border of the Syrian Desert. This chain of events may explain not only the departure of Abram and his clan from their paternal home and travel from Haran to the Land of Canaan, with the final goal of reaching Egypt. In the Bible this movement is explained as a result of a Divine Order, but we have to remember that the first step was that of Terah, about which we are told "And Terah took Abram his son….. etc……and they went forth with them from Ur of the Chaldees, to go into the land of Canaan etc." (Genesis 11: 31). More or less in the same period i.e. during the great dryness between the years 2300–1800 B.C., at ca. 1620 B.C. the invasion of Egypt by Semites also occurred, i.e. the Hyksos and they gained control over northern Egypt. Another result of the migration from the north towards Egypt may have been the establishment of city states of various nationalities as evidenced in the story of the conquest of the Land of Canaan by the Israelites.

Further on in the story of the patriarchs, the negative affects of the dry climate will be mentioned as a reason for their descending to Egypt. But when it comes to the beginning of the journey southward climate is not mentioned. In general, albeit the fact that the Middle East is a semi-arid region, one does not find many references in the Bible to climate variations. These are mentioned only a few times, as for example in the case of Noah's Deluge, The Ten Plagues, or when the people of Israel are warned about the rewards or punishments, which will follow the obedience or disobedience of the commandments of their Lord. The rewards will "be rain in due season, and the land shall yield her increase, and the trees of the field shall yield their fruit" and vice versa. Yet the climatic variations, which can be learnt from the various sources already discussed, lead to the conclusion that the decision of Abram-Abraham to move southward was in order to join the wave of immigrants from the plains south of the Taurus Mountains to Egypt.

An additional reason for the decision of Abram-Abraham to leave the homeland of his ancestors may have been his belief in an abstract god. One can even imagine that this brought him to try and convince his family and neighbors to accept his belief. As in all cases of a new idea, it can be assumed that his preaching to believe in a god, which can neither be seen nor touched was a source of friction between his faction and the adherents of the local gods carved from wood, stone or molded of clay. The fact is that after Abraham settled down in the Land of Canaan, and he looked for a wife for his son, he sent his servant to his family to bring the bride from his family. This may tell us that the religious gap between Abraham and his family back home was narrower than that between him and the local Canaanites. Not to speak on the fact that blood connections are stronger than beliefs.

In the Bible there is no description of the route of Abram-Abraham to Canaan. It can be assumed that he did not made a short cut southward as this would mean crossing a dry desert area, unless after a rainy winter. Taking into consideration that it was during a spell of dry years, a green desert seems to be a very low probability. It is more likely that he made a detour westwards, either through the mountainous backbone of the region or even along the foothills and the coastal plains.

It is possible that the clan of Terah, father of Abraham, belonged to one of the Semite tribes of the Mar-Yeminis, mentioned above. These tribes were roaming the area and from time to time were attacked by the kings of this region as described in the documents found in the archive of Mari. Once the climate became drier these wanderers headed southwards and westwards because the irrigated areas of Mesopotamia diminished and the population, local or foreign, were forced to leave by the inhabitants who had control of the lands and water resources, and who suffered from the drying up process.

The wave of the displaced population, including Hurrians and Amorites, on its way southward attacked the Canaanite city-kingdoms conquered a few of them. The conquerors were assimilated in the local population, adapted part of the gods and the local traditions forming the cultures of the small nations of biblical Canaan. The settlers built new cities according to a very well known and familiar North Syrian prototype like Carchemish and Mari, and turned them to fortified cities by surrounding them with thick walls with slippery slopes filled with compressed soil. Such are Tel-Dan, Hazor, Tel Beit Mirsim, Cabri, Shechem-Nablus, Ur-Shalem-Jerusalem, Lachish etc. When those who did not settle down continued their migration southward, penetrated into Egypt and took control over its northern part, the Egyptians referred to them as to hq' sq', meaning 'rulers of Foreign Lands' (Rainey, Notley 2006). The Greek transcription of the original Egyptian name is Hyksos.

According to the Bible's report on Abram-Abraham's journey southward, his first station in Canaan was at Shechem. This city is located on one of the principal roads of Canaan. It was and still is fed by a number of springs emerging from the limestone of Eocene age, building the mountains Ebal and Gerizim which rise west of the city. In our days the average annual amount of precipitation on these mountains is about 700 mm. Thus it can be assumed that also during a series of droughts, these springs will continue to flow. An acute climatic crisis might have caused their flow to reduce and even the drying up of a few of them, which may have caused an agricultural crisis in the area. The archaeological survey of Hirbet El Foka and Tel Balata, assumed to be the site of ancient Schechem, found pottery from the Early Bronze Period, which was a humid period, but no pottery was found from the Middle Bronze Period (2200–2000 B.C.) (Stern 2007). The meaning of this concurrence is that the settlements in this region were deserted because of the negative climate change, discussed above.

From the excavations in Tel Balata it was found that during the 18th century B.C., after the site has been resettled, the city was surrounded by a system of fortifications like those brought to the country by the conquerors from the north. These fortifications are surrounded by massive soil dikes on top of which was built a wall of bricks. In the case of Shechem the dike was covered from the outside with a strong wall built of great stones. It can be assumed that before these cities were settled and fortified the invaders from the north settled in the neighborhood in temporary habitations and only after they affirmed their rule, administrative as well as economic, did they started to build permanent settlements, fortifications and temples. Here again can be seen that the story of Abram-Abraham camping at Shechem or in the neighborhood is indeed an echo of the stories that passed from father to son about the migration of the Patriarchs who came to Canaan from the north following the wave of the conquerors.

According to the Bible, the stay at Shechem was a short one, and Abraham's clan continued the journey southwards and went down to Egypt. Yet, according to various data the hopes to find a prosperous Nile Valley were foiled. Various data show that between 2400 and 2000 B.C. the Nile floods were minimal bringing starvation

Fig. 8.1 The wall painting from Bani Hassan in Egypt describes a clan of Semites entering Egypt

to the local population, not to speak of the newcomers. The reason for this anomaly may be because of the rapid melting of the glaciers, which brought the cooling of the oceans although the climate was warmer. An echo of an economic crisis which also occurred in Egypt may be found in the story about the seven years of hunger, mentioned also in an Egyptian source. With the stabilization of the climate after this crisis, the equation of the warming of the climate and the increase in the amount of the summer rains in the tropic and the sub-tropic areas has fully recovered. This brought again a series of years of prosperity to Egypt, while Canaan suffered from dryness and hunger. This caused the wanderers arriving from the north to continue and reach Egypt. Thus the biblical stories that tell about the descents to Egypt reflect the climate changes around 2000 B.C., and the difference of the impacts of these changes on two neighboring countries located along the south-eastern coast of the Mediterranean, namely Canaan and Egypt. While the first suffered from dryness, the second enjoyed prosperity due to floods of the Nile.

The migration of clans of Semites to Egypt can be learnt also from different Egyptian sources. In a wall painting from a grave of an Egyptian clerk in Bani Hassan dated 1900 B.C., a group of Semites, men, women and children enter Egypt (Fig. 8.1).

References

Berlyn P (2005) The journey of Terah: to Ur-kasdim or Urkesh? Jewish Bibl Q 33(2):73

Buccellati G, Kelly-Buccellati M (1997) Urkesh: the first Hurrian capital. Bibl Archaeol 60(2):77–96

Green T (1992) The city of the moon god: religious traditions of harran. Brill, Leiden

Lloyd S (1978) The archaeology of mesopotamia. Thames and Hudson, London

Parrot A (1958) Mission archaeologique de Mari. Le Palais. Institut Francais d'Archaeologie de Beyrouth, Bibliotheque archaeologique et historique, vol 2. P. Geuthner, Paris

Parrot A (1968) Mission archaeologique de Mari. Le "tresor" d'Ur. Institut Francais d'Archaeologie de Beyrouth, Bibliotheque archaeologique et historique, vol. 4. P. Geuthner, Paris

Rainey AF, Notley RS (2006) The sacred bridge- carta's atlas of the biblical world. Certa, Jerusalem

Stern E (ed.) (2007) The New Encyclopedia for Archeological excavations in Israel, Israel Exploration Society and antiquities, Ministry of Defence-Publishing, Carta Jerusalem, Vol. Shchem

Weiss H et al (1993) The genesis and collapse of third millennium north mesopotamian civilization. Sci 261(5124):995–1004

Chapter 9
The Destruction of Sodom and Gomorrah

Abstract The abundant sources of water and the green pastures convinced Lot, the nephew of Abraham to choose the Dead Sea valley for his dwelling. He knew about neither the frequent earthquakes in this region due to its being a tectonic rift valley nor the morality of its inhabitants. The troubles of Lot started with the war of Mesopotamians kings including Amraphel, (probably Hammurabi) against the king of Sodom and his allies. The trouble continued because of a severe earthquake which destroyed Sodom and Gomorrah.

> Then the LORD rained upon Sodom and upon Gomorrah brimstone and fire from the LORD out of heaven;—And he overthrew those cities, and all the plain, and all the inhabitants of the cities, and that which grew upon the ground. (Genesis 19: 24–25).

The biblical story of the wandering of Abraham tells that after a short stay in Egypt he returned to the Negev and from there to the region between Bethel and Ai which is the mountainous part of Judea. Abraham's family included also Lot's family. As the still dry climate or the rate of births of the herds of both families was beyond the carrying capacity of the region, Lot decided to go down to the to the Dead Sea Rift Valley and plant his tent and his camp there, opposite Sodom: ***And the land was not able to bear them, that they might dwell together: for their substance was great, so that they could not dwell together. (Genesis 13: 6). And Lot lifted up his eyes, and beheld all the plain of Jordan, that it was well watered everywhere, before the LORD destroyed Sodom and Gomorrah, even as the garden of the LORD, like the land of Egypt, as thou comest unto Zoar.*** (Genesis 13: 10,11). We do not know the exact location of Sodom and Gomorrah that are mentioned in the Bible. For some reason it is accepted that the two cities were in the region of the southern part of the Dead Sea which today is an area of salt pans. According to the biblical story Lot decided to descend to the Dead Sea rift valley from the area of Bet-El, which is opposite the northern part of the Dead Sea it can be thus assumed that he saw the region of Jericho, watered by the springs of Elisa and Nuweima and the area east of the Jordan River, which is also watered by perennial springs. Like the oases north of the Dead Sea the oasis of A-Safi is also fed from springs emerging along faults. The water is infiltrating

A. S. Issar, *Strike the Rock and There Shall Come Water*, SpringerBriefs in Geography,
DOI: 10.1007/978-3-319-01937-6_9, © The Author(s) 2014

into the shallow layers of the alluvial deposits of gravel and sand, forming the alluvial fan of Wadi Hesi (Zered). Until quite recently this oasis was a jungle of canes, tamarisk and wild palm trees. Today the region is intensively cultivated .

Except the areas of Jericho and of E-Safi there are in the Dead Sea Rift, on its the eastern as well as on the western side, more sites in which springs emerge, because the rift constitutes a drainage base to the groundwater basins of both sides. Part of the water in these sites is fresh while part is brackish like part of the groundwater which outflows in the E-Safi and which arrives from a huge reservoir of ancient water that was collected in these layers during hundreds of thousands of years. Mount Sodom is a plug of salt built of these deposits (Fig. 9.1).

Thus the abundant sources of water and the green pastures convinced Lot to choose the Dead Sea rift valley for his dwelling. He did not investigate before he made this decision nor did he know about the earthquakes and their frequency, in this region. The troubles which overcame Lot, because of these two characteristics of the Dead Sea rift valley, may have caused him to be sorry about his decision but about this the Bible says nothing. Yet Lot's problems in the valley he had chosen to dwell in, started before the troubles inflicted

Fig. 9.1 Salt pillar—"Wife of Lot". Part of Mount Sodom, built of salt and clay layers. (Courtesy of Prof. Amos Frumkin HUJ)

later on the region because of its geology and because of the morality of its inhabitants (see Fig. 9.1).

> The Bible tells And it came to pass in the days of Amraphel king of Shinar, Arioch king of Ellasar, Chedorlaomer king of Elam, and Tidal king of nations;that these made war with Bera king of Sodom, and with Birsha king of Gomorrah, Shinab king of Admah, and Shemeber king of Zeboiim, and the king of Bela, which is Zoar. All these were joined together in the vale of Siddim, which is the salt sea. (Genesis 14: 1–3).

The personal acquaintance of the author of the present book with the sites described in this story started with his investigations of the water resources of the southwestern part of the valley of the Dead Sea for the Israeli Potash Company. The acquaintance with these sites continued when asked by the Iranian government, still ruled by the Shah-en-Shah (King of Kings) to visit the city of Elam, in eastern Iran, whose inhabitants complained of lack of water. In order to reach Elam the shortest road was the longest. As the travel from Kermanshah the capital of the province meant a rather difficult and dangerous journey, because of the steepness of the Zagros Mountains, the narrow mountain passes and the poverty of the local nomad tribes, which prompt some desperate people to ambush and rob travelers. Thus, a detour road was chosen, which meant going down from Kermanshah to the valley of the Tigris and traveling along the Iraqi border, including some distance inside Iraq, and then taking the ancient historical road which leads to Elam. To take this longer road involved also some dangers for an Israeli as Saddam Hussein, at that time the dictator of Iraq, wouldn't have liked it. Yet, holding a Laissez Passer of the UN could have helped when encountering Iraqi police. In addition to being obliged to make this journey because of the duties as a consultant to UNESCO to help the local people and their herds against suffering from thirst, this mission was readily agreed upon because it was to the city of Elam. This name Elam, was familiar since primary school, when the Bible was learnt and the story about Chedorlaomer King of Elam who joined Amraphel King of Shinar and the four kings led a 'blitz war' against the other kings of the Middle East. As a matter of fact, it was one of the first war stories to be read by the author in his youth; a few years later came the stories of World War II, the War of Independence and the wars which followed. The memory of the campaign of Chedorlaomer king of Elam and his allies against the king of Sodom and his allies, returned during the sixties of the late century, with the research for water resources for the Potash Company and thus the need to visit once and again ***the vale of Siddim, which is the salt sea*** (Genesis 14: 3), namely "the valley of lime, which is the Dead Sea". When I worked in this area the landscape was of a white plane of lime intermixed with saline water bodies formed, by the works of the Potash Company to dry up the southern part of the Dead Sea to turn it to evaporation pans (Fig. 1.3).

Also the story following the defeat of the king of Sodom and his allies was found to suit the environmental conditions: ***And the vale of Siddim was full of slimepits; and the kings of Sodom and Gomorrah fled, and fell there;***(Genesis 14: 10). In the present day, because of the above mentioned work on all the southern shallow part of the Dead Sea, it has become a plain of layers of white slime composed of gypsum and salt and lime. The defeated local kings escaping over this plane fell into wells (*be'erot* in Hebrew in the Bible) excavated in this layer to dig pitch (*hemar* in Hebrew in the Bible) as these wells have been dug for exploiting the pitch, which got buried in the sediments of the dried up southern part of the Dead Sea. Such blocks of pitch, emerge from time to time from the pitch layers found along the faults bordering the Dead Sea to float on it. Such blocks, which still appear from time to time, sank into the lime layers when the southern part dried up. The ancient inhabitants dug these blocks and marketed it via the city of Arad to Egypt, to be used there for embalming.

Dr. Yaakov Nir (2007) suggests that the pits in the "Sidim Valley" were sinkholes, similar to those which were formed along the shores of the Dead Sea, which were exposed after the lake's level dropped. This explanation is compatible on one hand with the climate picture, namely that it was a warm and dry period, which caused the level of the Dead Sea to fall and to the drying up of its southern part. On the other hand, the word "well" in the Bible means a hole in the ground dug in order to draw water as for example: ***And Abraham reproved Abimelech because of a well of water, which Abimelech's servants had violently taken away*** (Genesis 25: 21), or in this case pitch. It seems thus that the translation of *be'erot hemar* to *slimepits* is not totally correct, and should have been *pitch wells.*

Also the story of the series of war campaigns of the Mesopotamian kings against the kingdoms of the Levant may have roots in an historical event. Thus biblical Amraphel may be Hammurabi or Hamurphi, who ruled also over Shinar-Sumer, while the kings of Elam, as well as Elsar, maybe Larsa, which were under Hamurabi's dominion, participated in the invasion of Levant. Moreover, the dating of Abraham's times, namely during the first two centuries of the second millennium B.C. does fit with that of Hammurabi's, while some mixture of the events may have taken place, while the stories passed from mouth to ear.

The more general conclusion to be drawn from this story is that here again there is a distant echo on the dimension of time, which was passed from one generation to another, and which still preserved historical events like the war campaign of the kings from the east and views of ancient landscapes like the Sidim Valley a result of the drying up of the southern basin of the Dead Sea and on the wells for digging pitch, into which ones who are escaping may fall.

We return now to the stories of Lot's descent to the Jordan Valley, due to it being well watered. The reason for this statement is that these stories reflect natural events, which are due to the special geological structure of the Jordan-Dead Sea Rift Valley. In the first place, the fact that when Lot lifted his eyes he saw a land ***as the garden of the LORD, like the land of Egypt, as thou comest unto Zoar*** is because of the Jordan Valley being a down faulted rift valley, bordered on both sides by fault lines. These cause the water bearing strata to be exposed. Thus rainwater recharging these strata along the summits of the mountainous regions on both sides, flows as groundwater towards the rift and being exposed by the faults along the feet of the rocky walls bordering the rift flow out as springs. The story of the destruction of Sodom and Gemorrah also echoes historical events in the remote past. Indeed, the faults bordering the rift remain active to this day, and the movement along them may still cause destructive earthquakes. It goes without saying that in the Bible the destruction of Sodom and Gomorrah was not explained in geological terms, but neither was it explained by the failure of it inhabitants to properly worship their gods or pay them respect and proper sacrifices. Like in the story of Noah's Deluge, the biblical explanation is that this natural environmental disaster happened because of the breaking of the basic moral code of their society which brings upon them the wrath of the Divine Giver of Law. That these laws are still disobeyed can be learnt from the fact that this region is still undergoing earthquakes. If Sodom and Gomorrah were located in the south of the Dead Sea then the most suitable place to fit part of the description of their green environment is the area of E-Safi. Although this area cannot be viewed from Beth El, Lot may have moved south, for one reason or another.

As mentioned, the author of the present book got acquainted with the special characteristics of this region during the early sixties of the 20th century, when he was investigating to locate water resources for the Potash Industry in the southwestern part of the Dead Sea Rift Valley and later, during the seventies to locate sources for mineral water for spa sites, owned by Kibbutz Ein Gedi. The research project included the mapping, sampling and chemical analysis of all sites of emergence of springs. The water of several was of a high temperature and the contained bi-hydrogen sulphur gas (H_2S). Deposits of asphalt were also located. An oil exploration well drilled at one of these outcrops encountered saline water containing flammable CH_4 (methane gas).

These observations recalled the biblical stories about the catastrophes, which happened in the region where Sodom and Gomorrah were supposed to exist. It can be assumed that when a strong earth-quake has affected the area, the fracturing of the rocks along the fault lines bordering the valley may have opened an outlet for the saline water, which is under pressure in the depth. The emerging hot mineral water will be connected with the eruption of hydrogen sulfide and methane gas, which caused fires, suffocation and an increase of hot and salted geysers that flooded the fields and the gardens, and salted them. Additionally there was an emergence of hydrogen sulfite (H_2S) gas with inflammable methane gas. The falling rocks from the steep cliffs which rise above the rift add to the ferocity of the event described in the Bible.

Some people will find confirming evidence for the biblical story of the destruction of Sodom and Gemorrah in the pillar of salt, 20 m high, which people with creative imagination see as a woman (Fig. 9.1). These people and others claim that it is Lot's wife, who: ***looked back from behind him, and she became a pillar of salt*** (Genesis 19: 26). Professor Amos Frumkin (1991), who investigated the geology of Mount Sodom including this pillar of salt, came to the conclusion that this pillar is a part of the still rising Sedom salt diapir. It is a remnant of the wall of a cave formed as a solution channel in the salt. The walls of the cave collapsed due to a catastrophic earthquake, which happened about 1500 to 2000 B.C.

Possible sites where the sinning cities may have stood were suggested by archaeologists who worked in this region. The most outstanding results are from Bâb edh-Dhrâ, where Paul and Nancy Lapp (1974) discovered an Early Bronze Age cemetery site from the latter quarter of the fourth to the end of the third millennium B.C. and the continuation of the excavations by Rast and Schaub (2003).

At this site ashy soil and charcoal could be found on the surface. At other sites like Numeira two meters of dark ash and at Feifa evidence of destruction was found. The outstanding archaeological conclusion that all these sites were destroyed around the end of the Early Bronze III Age 2600–2300 B.C., while the period of the patriarchs is estimated to be between 1800 to 1000 B.C. namely Middle to end of Late Bronze. Thus, with all due respect to the archaeological research, no unequivocal evidence was found regarding the location of these cities.

For a general understanding of the geological background for the episodes of earthquakes and other phenomena peculiar to this region the reader is referred to the book *The Destruction of Sodom, Gomorrah, and Jericho* (Neev and Emery 1995).

To conclude the chapter regarding the destruction of the five cities around the Dead Sea, there is no claim that the geological and hydrological conditions in the southern part of the rift of the Dead Sea prove that the story about the destruction of the cities in the Valley of Sodom indeed happened as it is described in the Bible. Yet, it is claimed that knowing the natural environmental conditions in this region and the cultural background of this group of wanderers brought the present author to integrate the biblical stories in the environmental background, while taking into consideration the moral-theological worldview of the scribes who put these stories into writing. Thus while the conclusion of the hydro-geologist reading these biblical stories is that indeed during the period in which the patriarchs entered the land of Canaan, and maybe even before, there occurred a serious of earthquakes that destroyed some of the cities and villages in this area. The code of the ancient Hebrews explained the destruction and ruination of the cities and villages in the moral context, and not for example, as a failure to raise suitable sacrifices to the gods which will be palatable to their mouths and fragrant to their noses. From this point of view it is suggested to read again the part in which the Bible refers to the reasons that brought the destruction of Sodom : ***And the LORD said, Shall I hide from Abraham that thing which I do; Seeing that Abraham shall surely become a great and mighty nation, and all the nations of the earth shall be blessed in him? For I know him, that he will command his children and his household after him, and they shall keep the way of the LORD, to do justice and judgment; that the LORD may bring upon Abraham that which he hath spoken of him. And the LORD said, Because the cry of Sodom and Gomorrah is great, and because their sin is very grievous;*** (Genesis 18: 17–20).

In these verses the Bible presents Abraham as the one who has to guard the social moral laws and not as the guardians of ritual commandments of sacrifices and offerings. The emphasis is on: *and they shall keep the way of the LORD, to do justice and judgement*, which is the contrary to the sins of the people of Sodom.

References

Frumkin A, Magaritz M, Carmi I, Zak I (1991) The Holocene Climatic Record of the Salt Caves of Mount Sedom Israel. Holocene 1(3):191–200

Lapp PW, Nancy L (1974) Discoveries in the Wadi Ed-Daliyeh. In: Lapp PW, NL (eds), vol. XLI. Annual of the American Schools of Oriental Research, Cambridge

Neev D, Emery KO (1995) The destruction of sodom, gomorrah, and jericho: geological, climatological, and archaeological background. Oxford University Press, New York

Nir Y (2007) Was the Dead Sea level low like the current at the period of the war between the four kings against the five kings as it is described in the bible? Summaries and the program of the convention, the Israeli Association for Marine Science, the fourth convention, Haifa University

Rast WE, Schaub RT (2003) 1975–1981 Excavations at the town site of Babe dh-Dhra. University Press, Fribourg

Chapter 10
Abraham and Isaac Explore the Groundwater Resources of the Negev

Abstract The stories about the difficulties which the Patriarchs' herdsmen experienced in locating wells for the supply of drinkable water for their herds, describe the special hydro-geological conditions of the region of Beer-Sheva. The bedrock of this region is chalk, which is usually impermeable, except of limited areas where it is fractured. Water which penetrates these fractures dissolves and widens them creating subsurface channels enabling the flow of non-saline groundwater. Thus wells which happened to penetrate into such channels got a relatively ample supply of good water.

After his return from Egypt Abraham dwelt for a period of time in the mountainous part of Canaan which gets enough rain to water the pastures, albeit the land for grazing is rather limited. Thus as the climate became more humid Abraham went down to the Negev where in good years the grazing land is spacious while the water resources are limited. And after the day's grazing the herds gather around the springs or wells to get their water.

The Bible tells that on his first stay in the Negev. Abraham dwelled in the Gerar District: ***And Abraham journeyed from thence toward the south country, and dwelled between Kadesh and Shur, and sojourned in Gerar*** (Genesis 20, 1). Shur, later described as the Wilderness of Shur, is the northern part of the Sinai desert extending from what is today the Suez Canal to El Arish to the Negev, while Kadesh is either Ein Kadis or more probably the oasis of Ein-Kudeirat identified as Kadesh-Barneah, a valley irrigated by a few perennial springs, which had an important role in the story of Exodus, and which will be described and discussed in more detail in due course (Fig. 10.1).

All these sites either extend along the borderline between the sown land and the desert or are oases south of this border. Beer Sheva itself and the Tels, namely ancient towns or centers of rural population of this region, are located along riverbeds, like Wadi Beer Sheva, Besor, El Arish etc. The reason being, that the bedrock in this region is mostly impervious chalk. Thus the only water-bearing strata are either limestone rocks of Middle Eocene age, like

A. S. Issar, *Strike the Rock and There Shall Come Water*, SpringerBriefs in Geography, DOI: 10.1007/978-3-319-01937-6_10, © The Author(s) 2014

Fig. 10.1 The spring of Ein Kudeirat—Kadesh Barnea, Sinai (Photo by A. S. Issar 1956)

that feeding Ein Kudeirat or the sand and gravel layers in the riverbeds and solution channels along fractures in the chalks. These solution channels along fractures were formed in the close vicinity of the riverbeds by the flowing water. The local people utilized the stored water by digging wells to the layers of sand and gravel, and even deepen them into the chalk layers, once these layers were found to contain fractures and solution channels. As the water-bearing layers are recharged by flash floods, which occur only during the winter and only after a certain amount of rain on the river's drainage basin. If the quantity of rain is insufficient there will be no floods or only meager flows, the gravel and sand layers will not be recharged, the groundwater table will descend, and after a few dry years the wells will dry up. In general the gravel and sand layers underlying riverbeds with an extensive drainage basin will be thick and thus contain more water and the wells dug into these layers may ensure a supply of water even in dry years.

Thus when Abraham went down from the Hebron Mountains to the Negev he has chosen to dwell in the area of Gerar where a few streams which drain the south-western part of the Hebron Mountains and western Negev converge and continue as one riverbed to flowing to the Mediterranean. At present the underflow in

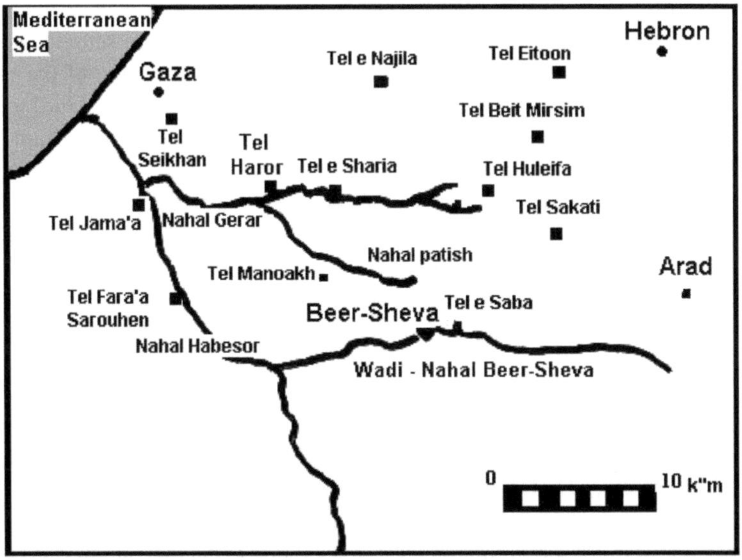

Fig. 10.2 Map of the region of ancient Beer-Sheva (Prepared by A. S. Issar, based on various sources)

these riverbeds appears on the surface to feed seepages and springs and to maintain perennial vegetation, and this river bed has been declared a National Park.

Yet the water in the river bed is brackish because it flows through the chalk layers which contain salt making it unsuitable for irrigation but good enough for watering herds. In good years the plains, covered by loess, are well watered by the rains and are ploughed and, while still moist, are sown, mainly with barley. After the harvest the Bedouins' herds are allowed to feed on the stubble. Thus the environmental and economical conditions of the present fit, in general, those which are presented in the Bible.

Along the embankments of the riverbeds crossing the region rise mounds or Tels which were inhabited in the Middle Bronze Period, the period when the patriarchs were living in this region. There is no archaeological evidence indicating on which of the mounds in the area between Beer-Sheva and the sea the city of Gerar was located. It was built like all the rest of the cities in this area from bricks of mud. The identification of the place is controversial; the most likely site is Tel Haror (Tel Abu-Hureira) (Fig. 10.2).

While the stay of the Patriarchs in the Land of Gerar was tolerated by the local inhabitants for a certain fee which included also the "lending" of their "sisters" to the local kings, it turns out that the biblical stories about the quarrels between the shepherds of the Patriarchs and the local shepherds took place in the eastern part of the Land of Gerar, in the place where the city of Beer Sheba is located. The ancient mound of Beer Sheba, located near the banks of the Wadi- Nahal Beer-Sheva which drains the southern part of the Judean Mountains and the northern plains of the Negev approaching the Mediterranean, was settled only from the Iron

Period. In a distance of a few km. from the mound there are ancient settlements of the Chalcolithic Period which was a cold and rainy period. The settlements were deserted at the end of this period as a result of an extreme change of the climate from cold humid to warm dry thus drying up of the water resources. Until today in the present mound of Beer Sheba (Tel e-saba) no remains have been found prior to the Iron Period, except for a few pieces of pottery of the Chalcolithic Period. It can be assumed that during the Bronze Periods, especially during the Mid Bronze Period, which most probably was also the period when the Patriarchs arrived and stayed there, no permanent settlement existed on the mound. Thus, it can be assumed that the biblical stories about the digging of wells at Beer-Sheva refer to one or several of the temporary settlements of nomads along the riverbed close to the mound of Beer Sheva. Even locating wells in the riverbed, does not ensure the finding of water, because of the special geological conditions in this region which will be discussed later.

The detailed Biblical stories about these wells tell of the conflicts and treaties between Abraham and later Isaac and the king or kings of Gerar. Though not historically precise, these stories also reveal the difficulty of locating wells for drinkable water in the environment of the riverbeds of Gerar and Beer-Sheva. In the case of Abraham the biblical story tells that: ***Abraham reproved Abimelech because of a well of water, which Abimelech's servants had violently taken away*** (Genesis 21: 25). In the case of Isaac the number of quarrels is greater:

> And Isaac's servants digged in the valley, and found there a well of springing water. And the herdmen of Gerar did strive with Isaac's herdmen, saying, The water is ours: and he called the name of the well Esek because they strove with him And they digged another well, and strove for that also: and he called the name of it Sitnah. And he removed from thence, and digged another well and for that they strove not: and he called the name of it Rehoboth and he said, For now the LORD hath made room for us, and we shall be fruitful in the land. And he went up from thence to Beersheba (Genesis 26:19–23).

For the author of the present book these stories reflect his personal experience when he was assigned to explore the potential supply of groundwater for the city of Beer Sheva. It was in the middle of the fifties of the twentieth century, when his professor Leo Picard was asked to help solve the acute problem of drinkable water supply to the city of Beer-Sheva. This city went through a rapid process of development when many new immigrants settled there. As an assistant the author carried out a survey of the wells which until then supplied water to the city. This survey and the research that followed shed light on the problems that Abraham and his son Isaac faced, and the cause for the quarrels between them and Abimelech's shepherds.

In the first place the geo-electrical survey of the riverbeds to locate sites of thicker layers of water-bearing sands and gravel showed that the subsurface river channels containing thicker layers of gravel and thus water are winding and not always continuous. Thus, without geo-electric equipment locating a well and encountering a gravel layer thick enough to supply the water needed for a herd, not to speak for a few herds, was rather a question of luck. The research also showed that although they are usually impermeable, over limited areas the chalks are fractured due to some pressure. Where these fractures happened to be under or along the river bed, the water penetrated and dissolved part of the chalk thus widening the fractures and creating subsurface channels enabling the flow of groundwater. Thus wells which happened to penetrate into such channels got relatively ample of water. Moreover the water of these wells was drinkable. On the other hand, a well which missed the channels and remained in the chalks was saline. Also the wells dug far away from the river bed may have encountered fractures but the water was saline because the chalk, originally deposited in the sea, still contained salt and, as it was covered by a thick layer of loess soil which is not permeable, and thus did not enable the rain water to penetrate into the fractures to wash away the salts.

From the story about Isaac's herdsmen, the emphasis put on finding 'springing water' (in Hebrew 'living water') namely fresh water, it can be guessed that they found a method to locate wells in the gravel or chalks along the banks of the riverbed and get plenty of fresh water. This brought the envy of the herdsmen of Abimelech and to their striving for Issac's wells instead of reaching some kind of an agreement.

Thus, the acquaintance with the geology and groundwater problems of the region of Beer-Sheva, the reading of the stories of the Bible herewith cited, either the quarrels between the herdsmen or the relations between Abimelech the king of the Philistines and Abraham and Isaac, leads again to the conclusion that:

a. Indeed these stories are based on incidents that did happen, but got mixed up and even distorted, while being transmitted from one generation to the next.
b. The geographical background of the stories is, indeed, that of the north-western Negev during a period of moderate climate, typical of our days and most probably also of the Middle Bronze Period. This moderate climate enabled not only pasture but even sowing in a more humid year: ***Then Isaac sowed in that land, and received in the same year a hundred fold*** (Genesis 26:12).
c. The distortion due to verbal transmission and the distance on the dimension of time is especially noticeable when it comes to alien personalities. In this case "Abimelech King of Philistines" has a Semitic if not Hebrew name meaning "king's father" or "my father is a king". The Philistines, who originated from the area of the Aegean Sea, settled in this area during the Iron Age, namely around 1200 B.C., more or less at the time when parts of Canaan were conquered by the Hebrews.

Some Bible scholars claim that the reason for the emphasis the Bible puts on the patriarchs' adventures in the Negev is because the scribes, who wrote these stories down, belonged to the Kingdom of Judea and their purpose was to show that the Hebrews had earned concession on this region from the beginning of their history. Yet, these stories demonstrate knowledge of the nature of the groundwater resources of the Negev. It is unreasonable to assume that the above-mentioned scribes went to the Negev to investigate the water resources in order to integrate them into the stories they wrote down. It is more logical to assume that knowledge of groundwater resources of the Negev came down along the dimension of time from those who learnt it from personal experience.

In the stories of the exploits of the patriarchs and their herdsmen, the reasons for the name of Beer Sheva are given twice. The first time because of the covenant between Abraham and Abimelech*: **Wherefore he called that place Beersheba because there they sware both of them*** (Genesis 21:31).

On the second occasion Isaac names the place, again to commemorate the covenant with the king of Gerar:

> And they rose up betimes in the morning, and sware one to another: and Isaac sent them away, and they departed from him in peace. And it came to pass the same day, that Isaac's servants came and told him concerning the well which they had digged, and said unto him, We have found water. And he called it Shebah: therefore the name of the city is Beersheba unto this day (Genesis 26, 31–33).

To swear or make an oath in Hebrew is *L'ishava* while 'seven' in Hebrew is *sheva.*

Investigation of the water resources of Beer-Sheva led the present author to doubt whether these explanations for the name of this city are founded on historical facts or are rather explanation given many years later.

As related in the former chapters, the archaeological excavations at Tel Beer Sheva have revealed that people lived on the mound ('tel' in Hebrew) already during the Chalcolithic period, around 4000 B.C. and it was fortified from the early Iron Age around 1000 B.C. or Israelite period. During this period of time the settlement on Tel Beer Sheva was enclosed by the outer back walls of the buildings which were built in a circle and whose entrance was protected by two towers. Outside the enclosed circle near the path leading down from the top of the mound to the riverbed, a deep well was dug. The time of the excavation is not known for certain, but the fact that its excavation did not disturb the hamlets and pits dating to Iron Age I speaks for its very early excavation. The location of the well is on the top of the hill about 25 m above the riverbed. The diggers of the well, most probably aimed to reach the groundwater table in the river gravel, but instead struck the semi-impermeable chalk layers, which did not deliver water. In the hope to reach water the diggers continued down to the depth of 57 m, water was located but not enough, so the digging continued to the depth of 70 m, where water was struck. The archaeologists Aharoni and Herzog who excavated the site suggest that the location of the well is connected with the religious cult of the people of this site (cited in Tsuk 2011).

In the opinion of the present author this suggestion is supported by hydrogeological considerations. Indeed the well, on one hand is not located on the alluvial terrace of the riverbed, which would have pointed to the intention of the diggers to ensure the water supply of the settlement, yet, on the other hand it is not located inside the gates, to ensure a supply of water in time of war. The conclusion then is that the well is not for water supply but is connected with the name of the site, namely the number 'Seven' or 'Sheva' in Hebrew, which had a religious meaning even before the time of the Hebrews. It is claimed that the holiness of this number was connected with the cult of the moon and with the counting of the days of the month, which begins and ends with the phases of the moon. The month was thus divided into four seven-day weeks, when the first night of the new moon and the last night of the old moon were days of festivity as well as the night of the full moon.

That the number seven had a religious meaning even before the time of the Hebrews is also evidenced from other excavations in Israel, as for instance from Naharia in northern Israel, where the archaeologist Dothan found a cult center of the Mid-Canaanite period (Mid-Bronze, ca. 2000 B.C.). Offering vessels made of seven cups were found near the local altar.

At a site from the Middle Bronze age, excavated near Tel Beit Shemesh, in the the holy of holies of the shrine, a candlestick with seven branches was found. It is thus suggested that the sacred significance of the number seven is connected with the cult of the moon and its fertility diety.

That the sacred significance of the number seven is connected with the lunar cult can be learnt from the second chapter in the fifth tablet of *Enuma Elish* which contained seven tablets, which dictates the duties of the gods and entrusts the night to be governed by the Moon-God:

"The Moon-God he caused to shine forth, the night he entrusted to him, he appointed him, a being of night, to determine the days. Every month without ceasing with the crown he covered him saying: ***At the beginning thou shinest upon the land, thou commandest the horns to determine six days and on the seventh day to divide the crown. On the fourteenth day thou shalt stand opposite the sun***" (King 1902).

In the Babylonian-Jewish calendar, the month was divided into 4 weeks of 7 days each. The Jews, though adopting the seventh day of every week as the holy day of Sabbath, starting on Friday night, detached it from the lunar cycle. The day of complete rest starts at sunset on Friday with lighting candles, as no fire can be kindled until the appearance of three stars after sunset on Saturday. On Friday evening special prayers are said of welcoming the 'Queen Sabbath' as well as during the day of Sabbath, while the phylacteries (tefillin), worn during every day morning prayers, are not worn. After the appearance of three stars after sunset at the end of Sabbath, a special hymn of parting from the Queen of Sabbath is chanted, a special candle is lit, and spices are smelled. Also, the Jews marked the first sighting of the new moon when it was at its minimum by a special prayers called "new moon blessing".

The relation between periodicity of fertility and the periodicity of the moon is manifested in those Hebrew's Jewish holidays, which have a connection with agriculture like Succoth which is the harvest festival and Passover, which is the spring festival and the harvest of the barley. These holidays start at full moon and last 7 days. The New Year of the Trees (*Rosh Hashana Lailanot*), which is the time of the first blossoming of the almond trees, telling about the beginning of the spring. Thus the 15th of Shevat became the date for calculating when the agricultural cycle began or ended for the purpose of biblical tithes is also at the full moon of the month of *Shevat*. Also the candlestick at the tabernacle and later temple of Jerusalem had seven branches.

To return to the patriarchs, the story of Jacob should be told now. Although this story, full of adventures, is outside the framework of the subject matter of this book, namely the impact of climate changes on history, the famine which struck next will be discussed.

As a preface comes the story of Joseph sold by his envious brothers, for twenty pieces of silver, to the Midianites merchants, who brought him into Egypt and sold him to Potiphar, an officer of Pharaoh, and captain of the guard. Next come

Joseph's adventures as a slave and later as a prisoner in Egypt. Then appears the story of the dreams of Pharaoh, in which the number seven plays an important role and displays Joseph's foresight, to first explaining the dreams and then advise how to cope with their impact.

> And the famine was over all the face of the earth: and Joseph opened all the storehouses, and sold unto the Egyptians and the famine waxed sore in the land of Egypt. And all countries came into Egypt to Joseph for to buy corn because that the famine was so sore in all lands (Genesis 41:57).

From the point of view of the regional climate regime, the Biblical report that ***the famine was so sore in all lands*** does not comply with the fact that the Nile belongs to a different climate regime than the lands neighboring Egypt. Thus, once there is a global warm climate and these lands will dry up, Egypt should enjoy a high Nile, due to increased rains falling on the subtropical region at sources of the Nile.

Thus either a global climatic anomaly took place, or as has been claimed a few times, the Biblical stories while based on true experiences of the ancients, were interpreted and changed according to the beliefs of each generation as they were passed down orally.

The core of the story may thus be that indeed a warm spell causing years of famine did strike the Middle East, which brought Jacob's clan to immigrate to Egypt, a step that his grandfather Abram took. Yet as this spell took a rather longer period than the previous one, the Hebrews settled down, but did not assimilate in their surroundings, neither in their faith nor in their occupation.

References

King LW (ed.) (1902) The Seven Tables of Creation, vol 1, p 274. Luzac, London

Tsuk T (2011) Water at the end of the tunnel—touring Israel's ancient water systems. Yad Ben Zvi, Jerusalem, pp 332–337 (In Hebrew)

Chapter 11
Egypt: A Land Like a Garden of Herbs

Abstract The patriarchs immigrated to Egypt as this country is watered by the perennial flow of the Nile, the sources of which are in subequatorial East Africa where the climate is humid tropical and subtropical. While the people of Egypt were of African origin the ruling classes might have been conquerors from Mesopotamia and the religion was thus an amalgamation of the natives' cults with that of the conquerors. The history of Egypt starts at the end of the Chalcolithic Period ca. 5200 B.P, when the lower and upper regions were unified, and hieroglyphics were developed. Later came the Early Dynastic period 3200 and 2686 B.C., during which the capital was established at Memphis, the cult center of the god Ptah. During the period of the Old Kingdom, which followed (Dynasties 3–6, from 2686 to 2181 B.C.) the first pyramids were built. During the period termed the First Intermediate Period (Dynasties 7–10, 2181–2050 B.C.) Egypt went through a stage of social and political strife. This helped foreign peoples, driven from their lands by famine, to penetrate and settle in Egypt. The period of the Middle Kingdom, which followed (Dynasties 11 and 12, 2050–1786 B.C.), was one of internal peace and prosperity. At the end of the Middle Kingdom, foreign rulers established themselves on the throne of the Pharaohs and ruled Egypt for about 100 years.

> The Land of Egypt where thou sowedst thy seed and wateredst it with thy foot as a garden of herbs. But the land whither ye go to possess it, is a land of hills and valleys and drinketh water of the rain of heaven. (Deut. 11:10, 11)

The aridization phase, mentioned already, which affected the region at about 4000 B.P. (2000 B.C.) at the end of the Early Bronze Age was characterized by the desertion of perennial settlement sites and the increase in the number of sites of nomads on the margins of the more humid regions. The wave of nomads who were not satisfied with the meager resources the desert could provide and did not have the courage or power to forcibly take the goods of the irrigated lands, moved along the borders of these lands trying to get something from the richness of these countries in exchange either for their products or for their services. In this context, one understands the many witnesses to the migration of Semitic tribes from the Fertile Crescent to Egypt one of which is the famous wall picture from an Egyptian tomb in Bani-Hassan already discussed (Fig. 8.1).

A. S. Issar, *Strike the Rock and There Shall Come Water*, SpringerBriefs in Geography,
DOI: 10.1007/978-3-319-01937-6_11, © The Author(s) 2014

The migration of the patriarchs to Egypt after they reached the Land of Canaan can thus be understood in the general framework of the movements of Semitic tribes from the semi-arid Fertile Crescent to its southern edge. Although this area is more arid it is supplied by the perennial water source of the River Nile.

It is thus interesting to note the fact that the two valleys, Mesopotamia as well as Egypt, each of which is more arid than most of the lands forming the Fertile Crescent, could serve as a firm basis to agricultural civilizations, and even supply a source of income to the refugees. This, as previously explained, is due to the fact that these two regions are crossed by big rivers, the sources of which are from rains falling on more humid terrains. Yet the stability of supply is a result not only of the abundance of rain in the drainage basin of the supplying rivers, but also of the storage capacity which is part of these river systems. In Mesopotamia the storage is supplied by the snows, which melt in spring, thus spreading the period of supply into the early summer, and subterranean space of the aquifers of the Anatolian and Zagros mountains. In the Nile system, the regulating storage is supplied by the lakes and marshes of subequatorial Eastern Africa. Another difference between the Mesopotamian and Egyptian river systems is the climatic belts to which they belong. While the Mesopotamian system belongs to the Mediterranean climatic belt, the Nile is fed by the tropical rains falling on east central Africa and the subtropical monsoonal system stretching north to the equator. The latter system is seasonally, as well as periodically, in an off-phase regime to that of the Mediterranean as was explained in former Chapters.

The Nile is the longest river in the world, measuring about 6600 km from its headwaters in Rwanda to its outlet in the Mediterranean Sea in Egypt.

Ever since man started his research into the mysteries of nature, the sources and regime of the Nile have posed an enigma. The Egyptians considered the Nile as one of their deities and saw in its perennial flow and seasonal fluctuation a mystery which should not be explained but worshipped. People with more inquisitive minds looked for a more natural answer, and one of these was Herodotus, a Greek who lived between 490 and 425 B.C. In his book 'The Histories' (1954) he wrote about *why the Nile behaves precisely as it does. I could get no information from the priests or anyone else. What I particularly wished to know was why the water begins to rise at the summer solstice, continues to do so for a 100 days and then falls again at the end of that period, so that it remains low throughout the winter until the summer solstice comes round again in the following year.* After failing to obtain a logical answer from the Egyptian priests, who most probably regarded this inquiry as blasphemy, and after arguing against the answers given by "certain Greeks hoping to advertise how clever they are", Herodotus gives his opinion. He attributes the fluctuations in the flowing of the Nile to seasonal difference between the lands on the shore of the Mediterranean which he knew and the lands south of Egypt from which waters of the Nile came. In his words, *…during winter the sun is driven out of his course by storms towards the upper parts of Libya. It stands to reason that the country nearest to and most directly under the sun should be most short of water. When the rough winter weather is over, the sun resumes its normal course in mid-heaven and from then on exercises an equal attraction over all rivers (namely, evaporates* *them) in winter, then all rivers but the Nile run in flood, because a great volume of rainwater is added to their volume. The Nile, on the other hand, behaves in the opposite way*. In a preceding chapter, after proving that the soils of Egypt are alluvial and were brought and deposited by the Nile, Herodotus states, *It seems to me therefore that if the land continues to increase at the same rate in height and extent, the Egyptians who live below Lake Moeris in the Delta and thereabouts will, if the Nile fails to flood, suffer permanently the same fate as they said would some day overtake the Greeks for when they learned that all Greece is watered by rain and not as Egypt is by the flooding of rivers, they remarked that the day would come when the Greeks would be sadly disappointed and starve*.

In this comment Herodotus has described the difference between the two systems of agriculture, the river-dependent system and the rain-dependent one. As already explained, the difference is the long-term storage capacity, which enables the big rivers to flow permanently, although fluctuations which are a function of seasonal variations cause seasonal changes in the flow of the rivers. On top of the seasonal variations, long term changes of climate may also cause a change in the rate of flow, which may cause catastrophes when the changes are extreme, by either flooding or by diminishing to famine level.

The relatively high constancy of the flow of the Nile as well as its periodical accurate fluctuations lies in the combination of several factors. The first is the climate which is dependent on the tropical system of the Trade Winds and not on the Westerlies which activate the Mediterranean system. For the explanation of this factor, which is also important for understanding other chapters in this book, some climatic background is needed.

The Nile is fed by rain falling on the subequatorial countries of East Africa, the source of moisture being southwesterly air streams from the Indian Ocean and the equator. The oceanic air blows inland in the summer, causing all the rivers north of the equator to flow. These summer rains affect mainly the Blue Nile and the Atabara rivers which originate from the north and central highlands of Ethiopia.

Thus, the floods of the Nile start in June and reach their peak in August. The White Nile on the other hand, is fed by rains falling on equatorial Uganda and southern Ethiopia, where there are two rainy seasons which results in more evenly distributed runoff. Another very important storage and regulating factor is the gigantic swamps of the Sudd into which the White Nile flows, and which cause the suspended silt of this river to be deposited and the supply of water to be regulated. In terms of relative contributions, the Blue Nile contributes about 60 % of the average annual water supply, the White Nile about 30 %, and the Atabara about 10 %.

Egypt, from the northern foothills of the Ethiopian highlands to the shores of the Mediterranean, lies in the desert belt girdling the subequatorial monsoon-trade-wind-affected belt to the north. The desert conditions are a result of the descending air masses which rise full of humidity over the equator, causing the heavy typical rains characterizing the equatorial belt. The descent of the air masses causes them to become hot and dry.

The annual precipitation over most of Egypt is less than 50 mm and only along the coast does it reaches 100 mm. The people of Egypt depend almost entirely on the Nile and it is no wonder that they regarded this river as a deity. The river itself was called Itern and its spirit was a deity called Hapi portrayed by a man with big breasts and a clump of papyrus on his head.

The Nile supplied not only water but also fertilizer as the floods of the Blue Nile and Atabara brought large quantities of silt which were spread over the fields by the flood waters. These black silts brought by the Nile were called "Keme" (The Black), while the silts brought from time to time by desert floods generated by Mediterranean storms were red and this desert silt was called "Deshret" (The Red). As we will see later, this difference in colors sheds some light on events recorded in the Bible.

The deity of the Nile was only one among eighty other deities which were worshipped by the ancient Egyptians. In the present study, we are primarily interested in the deities that are connected with the realm of water and enable us to understand the role of this realm on the history of the people living near and dependent on this water.

It is thus no wonder that the river was deified. The wonder is that it is not one of the most important gods for Egypt. The reason for this may be that when the people moved into the Nile Valley they brought their chief gods with them. As these primeval gods symbolized the powers of nature in their lands of origin, they simply added to their pantheon the local divine powers which became important only after they settled in the valley of the Nile.

Who were the people of Egypt and from where did they come? The most ancient artifact made by man in the earliest Stone Age came from eastern Africa. In the now arid parts of southern Ethiopia and northern Tanzania evidence has been found of the existence of a hominid tool producer more than a million years ago. This hominid went through several evolutionary stages, starting in the most Lower Paleolithic, continuing through the most Upper Paleolithic and into the Epi-Paleolithic period, which lasted until about 10,000 years B.P. During the last Ice Age (ca. 80,000–10,000 B.P.) the deserts of North Africa became more humid, which allowed a large population of hunters to spread over their plains, hunting the many animals which lived in the savanna-type environments that characterized many depressions in this vast country. At that time, subequatorial Africa, south of the Sahara became more arid, as can be seen from the low level of the lakes of this part of Africa.

The end of the last Ice Age brought about the aridization of the Saharan belt, so that the people who roamed the wide savanna-type plains which became arid had to search for a refuge. They migrated either south to subequatorial Africa, which became humid, or to the desert oases and the valleys, watered by either springs or rivers. Remains from the Neolithic Period are rather rare in Egypt. Some Late Neolithic relics were found in the Faiyum area. The emmer, barley and flax found there disclose connections with Asia, but the tools and weapons clearly indicate African traditions. During the Chalcolithic Period man had already settled at many places along the river. Various sites containing relics of typical cultural traditions have been excavated. The people of this period already believed in life after death, as they provided their dead with a great supply of objects to help them in their new form of “existence”. Cultural connections with other parts of the Middle East can be asserted from the female figurines with especially emphasized procreative attributes, symbols of a fertility cult.

While the bulk of the people of Egypt were the descendants of the prehistoric tribes that settled along the Nile when the desert areas first became dry, the ruling classes might have been conquerors from abroad, most probably from Mesopotamia.

At about 5400 B.P. (3400 B.C.), a drastic cultural change affected Egypt. The country was divided into two kingdoms, one of Upper Egypt, the other of the Delta region of Lower Egypt. At the same time the art of writing appeared as well as other arts typical for Egypt, such as their architecture, drawings and sculpture.

Various paintings of this period show a Mesopotamian influence. An excavated knife handle shows a sea battle between Egyptian ships and others which look Mesopotamian. Findings in graves show that people with a different anatomy (larger, with a differently shaped skull from the ordinary Egyptian), and using special funerary architecture, formed some kind of ruling aristocracy. These people were named also “Followers of Horus”. They were divided into ruling houses, each governing one of the kingdoms of Egypt.

At about 5200 B.P. (3200 B.C.), the two kingdoms were united into one by the King of Upper Egypt, Menes or Nar-mer. He is believed to have founded the town of Memphis, the original name of which was Het-Kau-Ptah, the House of Ptah which later was transformed into Aiguptos, hence Egypt. With King Nar-mer the First Dynasty began which, with the Second Dynasty, continued until 2686 B.C. During this period, also called the Archaic Period, the art of writing developed to the high level of hieroglyphics found in the pyramids.

The religion which developed at this time was the amalgamation of the native cults of the local tribes and those imported by the conquerors. The gods of the conquered were not suppressed but were annexed to the Pantheon, or even adopted as sons of the conquering deity. They remained chief gods in their original provinces, to be worshipped for millennia.

At the time of the unification, the rulers worshipped the sky god Horus, whose symbol was the falcon. Seth was the god of the original people. The replacement of Seth by Horus after a period of struggle is narrated in mythological tales of the triumph of Horus, symbolizing good, over Seth, representing evil.

During the second dynasty there came a fusion between the sky god Horus and the sun god Re and the king identified with Horus became the son of Re. A local god in the Delta area was Ptah, who was believed to be the creator of the universe. Another important deity originating in Lower Egypt was Osiris. He was believed to have been killed by Seth and avenged by his son Horus. The Pharaohs were worshipped as demi-gods, believed to be the descendants of Horus. This most probably symbolized the victory of the god of the conquering race over the local deity.

The worship of many deities which sometimes overlapped each other in their dominions characterized the ancient Egyptian religion and, in some ways, can be compared to the Hindu philosophy of religion which tolerates different answers to the same problem as different facets to the same phenomenon. For instance, the creation of the world was attributed at Memphis to Ptah. At Heliopolis, Re the sun god was believed to be the creator. At Elephantine it was believed that Chnum the god in the form of a ram was the creator (Herodotus 1954, Emery 1961, Frankfort 1948, Frankfort et al. 1946, Clark 1959).

The ancient controversy between the ram and the bull worshippers was solved by the Egyptians adopting also the bull—Apis, which symbolized strength in war and fertility. He was believed to be the manifestation of the god Ptah, the creator. His cult on a national basis was established most probably by Menes, the founder of the First Dynasty. Hathor was the sacred cow, patron of the skies, love, and joy.

Were these Gods remnants of the prehistorical cults of the people of the rams and the bulls? Most probably, yes. The Nile valley seems to have been a huge religious 'soup' to which every ancient cult became an added ingredient, but not the main component.

It is thus interesting to inquire whether the people coming from Mesopotamia at the end of the Chalcolithic Period to seize power and create the ruling class also brought with them their cults from their land. We heard that they were the people of Horus the hawk, a cult which was not recorded as such in the ancient Near East. On the other hand, we know that, since the First Dynasty, the worship of the sun had started to be a national religion. Horus, too, was worshipped as the god of the sky, and thus goodness, in contrast to Seth, who was most probably the god of the aborigines of the Valley of the Nile. He was represented by some unidentified animal with the appearance of a dog and the head of an ant eater.

It thus seems probable that the people who came from Mesopotamia or from another part of the Fertile Crescent were worshippers of the sun and, although they accepted the local gods as members in the club of the deities, they insisted on their being second class members and even threw out the former head, Seth, denouncing him as a symbol of evil.

From Egyptian mythology one can discover what could have been the state of the religions of the Chalcolithic tribes that invaded Egypt. We saw that they worshipped the sun, the hawk, and the bull. The bull and the sun were worshipped during the Neolithic and Chalcolithic in Anatolia, and even reached Canaan. One can thus assume that some time during the Chalcolithic the climatic conditions enabled the people from the north to reach the border of Egypt. When the climatic conditions worsened at the end of this period they looked for a refuge for themselves, their cattle, and their gods in the land with a perennial supply of water.

It can also be claimed that these people did not come from one of the well established, settled Mesopotamian cultures, as one can find references in the Egyptian mythology analogies only to rudimentary items characterizing the Mesopotamian religions. As will be shown later, most beliefs are lacking the story of the Deluge and if similar myths exist they are a very distant echo of something long forgotten.

Yet one can find in the Egyptian stories of creation some metaphysical ideas which are related to those found in the book of Genesis and the Mesopotamian myths of creation. The most prevalent is that of the Primeval Waters as the primary entity from which all the world was formed. This idea is common to all the Egyptian accounts of the origin of the universe. Every creation myth in the different Egyptian mythologies assumes that, before the beginning of things, the Primordial Abyss of waters was everywhere, stretching endlessly in all directions (Simpson 1972). The rising of the Creator from the Primeval Waters was also the manifestation of the creation of land, light, order, and life.

The "pre-creation" state, namely, when Atum was still immersed in the Primeval Waters was a state of unhappiness and pain. He was very weary and inert thus the waters represent helplessness and chaos. At the same time, immersion in them means going back to primeval innocence.

The Primeval Waters were represented by the deity Nun. In many texts, the description of the Primeval Waters is given in negative forms of speech, defining it by what it is not: *Where the Universal Lord dwelt when he was in the infinity, the nothingness, and the listlessness.* According to the cosmogony of the cult center of Hermopolis in Central Egypt the waters produced four beings out of themselves: (a) Nothing, (b) Inertness, (c) Infinity, and (d) Invisibility or Darkness. These were given female counterparts and were worshipped as demigods or goblins in the town of the "Eight Ones" or Shumunu (Eight), later called Hermopolis. They formed the Primeval Egg in the darkness of Father Nun. From this egg emerged the 'Bird of Light'. Other sources maintain that the 'Primeval Egg' coming from the Primeval Waters was filled with air which separated Geb, the earth, from Nut, the sky, a function usually attributed to Shu aided by the wind spirits. The act of separating Geb from Nut also has some vague similarity to the separation myths discussed previously, in the Mesopotamian and Biblical stories. In Memphis, the capital of the Old Kingdom, the high god worshipped was Ptah. He is described as floating in his reed boat across the sky-ocean, looking down upon his creation.

In a more abstract myth of creation, the Primeval Waters are symbolized by sleep and primordial chaos from which the living soul struggles to free itself. The high god rises from the primeval waters as a high hill on which Heliopolis as a cult center was built. The appearance of Atum, the creator, was as the Primeval Mound which was also the creation of light, since the waters were in absolute darkness. Atum, or "The Complete One", was afterwards worshipped as Re presented as the sun.

The worship of Atum-Re as high god flourished during the Old Kingdom (from ca. 2780 B.C. to ca. 2500 B.C.). This was the period in which the pyramids were built. The main religious center was at Heliopolis, (today a suburb of Cairo), the priests of which gained supreme power and had their cosmogony inscribed in the pyramid texts. The idea of Atum as the only high god went through a series of changes during the period of the Old Kingdom. One of its presentations is in the form of the beetle, *Scarabaeus sacer*. The insect, observed in the act of rolling his egg enclosed in a ball of dung, symbolized Atum as he came into existence as the rising sun. The different manifestations were explained in the following way. Atum is the aboriginal invisible god who is seen in his visible forms as Khorper, the scarab, and Re, the sun in the sky. Gradually Atum, who is invisible, becomes the night sun as it journeys through the underworld.

Another manifestation of the creator is that of a serpent. In the pyramid texts, one finds the following narration:

I am the outflow of the Primeval Flood, he who emerged from the waters.

I am the 'provider of Attributes', the serpent with its many coils.

The serpent, however, symbolizes not only the emergence of creation from the water but also the end of existence—"the great surviving serpent when all mankind has reverted to the slime". A conflict between Re, the sun god, and an enemy serpent is also encountered in the Heliopolis religion, the serpent appearing in the form of a curly-haired maiden, a legend in which one can find some relation to the story of the Snake and Eve in Paradise.

The Old Kingdom continued from the Third Dynasty, starting at ca. 2780 B.C., until the end of the Sixth Dynasty, at about 2250 B.C. The transition between the Old and the Middle Kingdoms was a period of rebellion and anarchy, yet intense literary activity flourished. In many of the remaining texts from this period, mainly the Coffin Texts, there is evidence of a more abstract way of grasping the acts of genesis and world order, but still the notion of the primeval waters as the beginning of everything prevails.

I was the Primeval Waters, he who had no companion when my name came into existence (Coffin Text 715, Clark 1959). While still in the water, the spirit of the creator becomes a circle which is replaced by the symbol of the cosmic egg later it turns into Hahn, which is the cosmic wind which divides and separates the sky and the earth. From the creator emerge Command and Intelligence who go round the circle of total being giving each thing its name, that is, its distinctive characteristic.

I am the Eternal Spirit, I am the sun that rose from the Primeval Waters, my soul is god, I am the creator of the word, evil is my abomination, I see it not. I am the creator of the order wherein I live. I am the word which will never be annihilated in this my name of Soul (Coffin Text Spell 307, Clark 1959).

In conclusion, it is the opinion of the present author that the Egyptian myth of the Primeval Waters does not differ greatly from that of the Primeval Waters symbolizing chaos in the Mesopotamian and Biblical myths, namely, "The primeval Apsu, Mummu and Tiamat, the waters of which were mingled together" of the Babylonian-Sumerian Enuma Elish and the "Darkness over the deep" 'Tehom' of the Hebrew Genesis, a similarity which points to distant yet common roots.

Another point of resemblance between the Egyptian religion and that of the Fertile Crescent, the origins of which the author believes have a common archaic basis is the cult of the god killed and resurrected, namely of a deity which follows the cycles of the seasons: death symbolizing the dryness of the summer, and resurrection, the sprouting of the new vegetation after the rains or floods. In the Mesopotamian and Canaanite mythologies, it is the death and resurrection of Tamuz—Dummuzi and Baal and in Egypt it is Osiris.

In the Jewish faith is in resurrection of the dead after the Messiah will come 'at the end of the days' and will blow his horn. The Christians believe in the first place in the resurrection of Jesus Christ and also in miracles performed by apostles and in the resurrection of dead after Judgment Day. A similar belief in the resurrection of dead after Judgment Day exists also in Islam.

In Egypt the belief in resurrection can be understood in light of the special environmental conditions of this fertile country surrounded by a desolate desert. The dry season just before the floods were due had the full meaning of death if the floods of the Nile failed to arrive. The floods cover all the land turning it into a vast lake, bordered on both sides by the desert. The water had to retreat before sprouting could start. All these events are symbolized in the myth of Osiris. It starts with a golden age when Osiris was the ruler of the land, and order and civilization were at their height. This was destroyed by Seth, his younger brother, who killed Osiris, mutilated his body and threw it into the Nile. Some legends tell of Osiris being put in a coffin and drowned in the Nile. Isis, the sister and wife of Osiris, (compare with the Mesopotamian Ishtar and the Canaanite Anath) helped by her sister, Nephtys finds the body, collects all its parts, brings it to land and revives Osiris. She bears him a son whom she hides in the swamps of the Delta. The son is Horus who grows up and takes vengeance for the evil done to his father by attacking Seth. The war between the gods is a long episode during which order is disturbed in the world. In the end the gods intervene. Arbitration by the god Thoth decides that Horus is the legitimate heir of his father, Osiris. Horus then travels to the underworld where his father still lies, to bring him the news. Osiris revives and becomes the spirit of new life and growth. Seth's punishment is to become the boat which carries Osiris on the Nile during the fertility festival.

The cult of Osiris went through many stages of evolution to the final stage in which Osiris was regarded as the chief deity of the Egyptian pantheon. The story of Osiris links with the myths of Mesopotamia and Canaan in the basic theme of the fight between the primordial powers of disorder and a new power which represents order and organization. Again one can see that, although the Egyptian religion developed and evolved independently of the other religions of the Levant, yet its roots bear common denominators with those of Mesopotamia and Canaan.

The author suggests that these roots stem from prehistoric times, primarily the Chalcolithic Period, before the Mesopotamian civilization developed its main characteristics. Thus, while one finds similarities in the basic themes of the archaic myths of creation and fertility, the story of the deluge, which was developed in Mesopotamian civilization in the preliterate period, is very different in the two cultures. In the Egyptian mythology, this story starts with Re asking Hathor—the Cow-Goddess, to destroy mankind because they plotted against him, but when she proceeds with this task he changes his mind. He orders servant girls to prepare an intoxicating red brew from barley and red ochre with which he floods the country to the depth of three palms. When Hathor comes at dawn to slaughter what remained of mankind, who meanwhile had fled to the desert, she sees her reflection in the water. She drinks the water, which is beer, becomes drunk, and forgets her task of annihilating mankind. After that the ritual preparation of beer by servant girls for Hathor was part of the festivities during the annual festival. As can be seen, there is very little in common between the two legends. While the Mesopotamians described the flood as a disaster, in Egyptian it was described as a savior. Thus, the present author's opinion is that the story of the Egyptian deluge is connected with a severe drought symbolized by the people fleeing to the desert. The drought was brought by Re symbolized by the sun. The same sun also brought the silt-laden floods of the Nile, which saved the people.

Thus, it might be that the Chalcolithic Period (approximately 5400 B.P.), when a big change influenced Egypt to become abruptly converted from a state of advanced Neolithic village cultures to two well-organized monarchies with many signs of Mesopotamian incursion, was also the time in which the branches of the archaic myths of creation and resurrection were grafted onto the local myths to later develop into the very special Egyptian civilization.

A short summary of the history of Egypt as an introduction to the next chapter will conclude the present one.

Some time at the end of the Chalcolithic Period (ca. 5200 B.P—3200 B.C.), the two parts of Egypt, lower and upper, were unified, and writing was developed into hieroglyphics.

The period between 3200 and 2686 B.C. is called the Early Dynastic period (Dynasty 1 and 2), during which the capital was established at Memphis, the cult center of Ptah.

During the period of the Old Kingdom, which followed (Dyn. 3–6, from 2686 to 2181 B.C.) the first pyramids were built, beginning with step pyramids and developing into the classical pyramids at the end of this period.

During the period termed the First Intermediate Period (Dyn. 7–10, 2181–2050 B.C.), Egypt went through a stage of social and political strife. This helped foreign peoples, driven from their lands by famine, to penetrate and settle in Egypt.

The period of the Middle Kingdom, which followed (Dyn. 11 and 12, 2050–1786 B.C.), was one of internal peace and prosperity. From this period there are connections of trade and culture with the Land of Canaan, known from archaeological findings in the two countries. At the end of the Middle Kingdom, foreign rulers established themselves on the throne of the Pharaohs and ruled Egypt for about 100 years.

References

Clark R (1959) Myth and Symbol in Ancient Egypt. Thames and Hudson, London
Emery WB (1961) Archaic Egypt. Pelican, Gretna
Frankfort H (1948) Ancient Egyptian Religion. Harper and Row, New York
Frankfort H, Wilson HA et al (1946) Before Philosophy. Pelican, Gretna
Herodotus (1954) The Histories. Translated by Abry de Selincourt. A penguin classic
Simpson WK (eds) (1972) The literature of Ancient Egypt, an anthology. Yale University Press, New Haven

Chapter 12
Foreign Rulers: The Hyksos. Foreign Climate: The Ten Plagues

Abstract A spell of a warm and arid climate at ca. 1800 B.C. brought the Hyksos invasion of Egypt and at about 1700 B.C. a king of Hyksos origin occupied the throne. At the same time officials of Apiru (Hebrew?) origin reached high posts in the government of Egypt enabling their clans still living in drought-stricken Canaan to enter Egypt and settle on irrigated land in the delta area in the land of Goshen. The Hyksos were overthrown in 1558 B.C. by the Egyptian Ahmose and the Apiru were enslaved. Toward the end of this dynasty king Akhenaten brought about a religious revolution by shifting the seniority in the Egyptian pantheon from Amun to Aten, the sun god. A cold humid climate, which started about 1300 B.C., had a negative environmental impact on Egypt, which forms the core of the story of the 'Ten Plagues'. The cold humid spell caused the greening of the Sinai Desert, which encouraged the Hebrews to leave Egypt.

> And Joseph said unto his brethren… For these two years hath the famine been in the land and yet there are five years in which there shall neither be ploughing nor harvest… and he made me a father to Pharaoh and lord of his house and a ruler throughout all the Land of Egypt (Genesis 45, 6–9)
> And Moses stretched forth his rod toward heaven: and the Lord sent thunder and hail, and the fire ran along upon the ground and the Lord rained hail upon the land of Egypt. (Exodus: 9, 23)

From the paleo-climate curves (Figs. 1.2 and 1.5) it can be learnt that at ca. 1800 B.C. there occurred a marked negative shift in the precipitation amount as registered by the stalagmites' rings of the Soreq Cave, a fall of the level of the Dead Sea and the humidity curve of Lake Van. This extreme drying of the Land of Canaan brought the foreigners who immigrated from the north to invade Egypt, which was independent on the rainfall regime of the Middle East but on that of the Tropics and Sub-Tropics. Moreover, the warm climatic spell, which badly affected the lands from which these foreigners came and stayed, had a positive effect on the water regime of the Nile and thus to the economy of the countries dependent

A. S. Issar, *Strike the Rock and There Shall Come Water*, SpringerBriefs in Geography, DOI: 10.1007/978-3-319-01937-6_12, © The Author(s) 2014

on this river, namely the Nile. It did not take a long time before the wanderers from the land of Canaan, among them the clan of the Patriarch Jacob left the countries hit by frequent droughts and thus famine and migrated to Egypt.

This happened in 1780 B.C. when Egypt underwent an extreme change in its regime. The rulers of the country were no longer Egyptians, but foreign people of special military talents who were called Hyksos. This name is derived from Helenistic sources while the original Egyptian name was like Haka-Hashut meaning "rulers of Foreign Lands". They came from the northern part of the Middle East. In the Land of Canaan they overwhelmed the local inhabitants, probably by their skill in mobile warfare, using light war chariots drawn by horses onto the battle field and siege machines to conquer the towns. Afterwards, they refortified these towns with huge earthen ramparts to protect the walls against siege machines.

The exact origin of the Hyksos is not yet known. The Egyptians refer to the raiders coming from the east, as "Apiru", which reminds of the name "Hebrews". Some of the kings of the Hyksos also had Hebrew sounding names. Opinions differ on whether the Hyksos were Hurrians, who came from the northern part of the Fertile Crescent, probably southeastern Anatolia, or were they Amorites who embraced Hurrian ways of war, or a multiracial migration from the northern part of the Fertile Crescent southward. The Hebrews most probably followed the hard core of warring tribes who seized power in the Canaanite-ruled cities and later moved southward into Egypt.

What could have caused this migration of people? We have seen that, from ca. 2200 to 2000 B.C. a period of dry years compelled the Amorite people to escape from the parched semi-arid lands into Mesopotamia where they replaced the ruling society of Sumer. At this time, which coincides with the First Intermediate Period in Egyptian history, we know from Egyptian sources that the infiltration of people from Canaan into Egypt had also begun. As one can see from the records, some of the newcomers to Egypt were welcomed by the Egyptian authorities probably because of their skill in some of the arts needed by the Egyptians. The picture from the tomb at Beni Hassan, described in the preceding chapters, especially of the bellows on the back of the donkey, surely tells something.

The historical records tell that from ca. 2200–2000 B.C., there were problems on the borders of Egypt, but these were more with the Libyans to the west. Libya is also a Mediterranean type arid zone, affected in the same way as the Fertile Crescent during the winter by moist air originating over the seas of the Northern Atlantic and Mediterranean. The Egyptians, who had enough food due to a high Nile, have allowed small groups of the hungry people crossing the desert to enter their country peacefully from the east. These newcomers could supply services needed by their rich society. On the other hand the Egyptians had no reason to share their food with the hungry nomads on their western border, whom they fought back.

The movement of the Hyksos down from the semi-desert of the Anatolian plateau into the coastal plains of the eastern Mediterranean most probably started with the drought that caused the great Amorite movement into Mesopotamia. Because of the same drought, some time afterwards the Hebrews may have started their wandering movement. As the western margins of Mesopotamia were at that time already occupied by the first wave of Amorite and not being able or not being willing to find a permanent place to settle, they followed the migratory trend southward with the wave of the Hyksos in the hope of finding a source of livelihood in the fertile coastal plain or the valleys of Canaan.

Some of these tribes were able to secure a holding in the more fertile valleys irrigated by perennial springs. The family of Lot, the son of Haran is an example of such settlement. In broader terms, the inhabitation of the more humid mountains of Trans-Jordan by the Semitic Ammonites, Moabites, and Edomites may reflect such a process. Thus the migration into Egypt can better be described as an act of infiltration and settlement by small groups, mainly in the Delta zone of northern Egypt.

The biblical stories of the migration of the Patriarchs between Canaan and Egypt reveals the character of this type of movement, which reflected the changes between dry and wet spells of a relatively arid climatic period. The later settling of.Jacob and his sons in Egypt most probably occurred when the climatic conditions became too arid in the Land of Canaan. From the more detailed biblical narration of these events, one hears some echoes of the environmental and socio-economic background which existed in these lands in the period of the settling of the Hebrews in Egypt, Abraham and later Isaac descend to Egypt and return to Canaan. Later, Isaac settles in the area of Beer Sheva, the semi-arid part of the Land of Canaan. He has several disputes with the local population who were called Philistines (after the name of inhabitants who will later occupy this region), but had Semitic names and were a part of the Canaanite tribes living in this area. We have already described the hydrological background of these disputes. Following a family feud, Isaac's son Jacob escapes to northwestern Mesopotamia, where he joins his former clan, marries, raises a family and then migrates back to the Land of Canaan. On his return through the Syrian Desert he encounters difficulties with his brother's tribe, the Edomites, but is able to reach an agreement. He then enters the Land of Canaan and moves with his tribe and herds between the fortified Canaanite towns, some of which might already have been occupied by Hyksos clans. At the town of Shechem, a feud, caused by the breaching of the code of honor by the local people, brought about the storming of the city by the sons of Jacob. To avoid vengeance, they moved quickly southward to the border of the desert, which was the natural habitat of nomad tribes. A spell of severe drought causes the family to request permission to enter Egypt. They are helped by a member of the clan who had reached Egypt before them and has risen in the local government to be the viceroy. This might have been due to the Hyksos and Apiru people's pervasion of the government of Egypt during the Intermediate Period..

Historical records show that at about 1700 B.C. the Hyksos gained full control over Egypt and a King of Hyksos origin occupied the throne. It is quite logical to assume that at first officials of Apiru origin reached high posts in the government of Egypt enabling their clans still living in drought-stricken Canaan to enter Egypt and settle on irrigated land in the Delta area. Most probably, these people were granted special rights and enjoyed their stay in this country. The province granted to the Hebrews to settle down was The Land of Goshen, Egypt located in the eastern Delta (Fig 12. 1), which was suitable for their livestock.

Fig. 12. 1 Map of the delta of the nile and western sinai. (based on satellite image by NASA)

The capital city of Egypt under the Hyksos was at Avaris (present Tel el Daba– The Hyena's Hill) (Fig 12. 1) located in the eastern part of the Delta of the Nile. It was destroyed in 1558 B.C. by Ahmose I, after he defeated and expelled the Hyksos. This led, apparently, to the Apiru being deprived of their special privileges and even being enslaved and forced to work in the royal construction projects.

While the migration from the Fertile Crescent and the gaining of control over Egypt, was due to a climatic crisis in the Canaan while Egypt enjoyed a high Nile, the downfall of the Hyksos might have been connected with a global cold climatic change, which caused an economic crisis in Egypt, but brought a more humid climate to the Fertile Crescent. Hunger and strife among the Egyptian farmers, on one hand, and the concentration of wealth and power by an alien ruling class, on the other, may have triggered the period of unrest which brought the overthrow of the hated dynasty in its wake. The revolt against the Hyksos was led by the two brothers, Kamose and Ahmose who finally defeated the foreigners and expelled the Hyksos. After finally ascending the throne in 1558 B.C., the Egyptian King Ahmose reorganized the country and the army. He and his descendants, who comprised the 18th Dynasty, even organized war missions into the Land of Canaan From the engraved memorials of the kings of this dynasty it can be learnt that in some way Canaan became a protectorate of Egypt.

Toward the end of this dynasty, one of the kings brought about a religious revolution by shifting the seniority in the Egyptian Pantheon from Amun to Atum or Aten, the Sun God. This king also changed his name to Akhenaten and moved the capital of Egypt southward to a place called today El-Amarna to get away from the influence of the clergy devoted to Amun. The archives of the court containing many clay tablets written by the kings of Canaan, Akkad, and Hittie were found in this new capital. As can be learnt from these tablets, most of the kings of Canaan regarded the King of Egypt as their protector and arbitrator in local feuds. The religious reform did not continue after the death of Akhenaten. The priests of Amun again gained control of the state's religion, re-crowned their god as chief, and removed all trace of the sun reform.

Some scholars maintain that the biblical Exodus took place during the reign of Akhenaten, who was occupied with religious matters. Some see in the sun religion some kind of a shift toward monotheism and compare it with that introduced by Moses. All scientists who believe that the Exodus of the Bible was a historical event, place it between the end of the Hyksos rule over Egypt at the beginning of the 18th Dynasty (1558 B.C.) and the reign of Merneptah of the 19th Dynasty (1220–1238 B.C.). The later date is based on the fact that in a stele of this king the term Israel is first mentioned, claiming that it was destroyed for eternity.

Although the last statement was undoubtedly premature, this document shows that the Israelis-Hebrews, were considered as a nation the king of Egypt has to fight, destroy, and boast about it.

The various opinions regarding the date of Exodus are discussed by Wilson (1985). Lately more voices connect the biblical Exodus with the volcanic explosion of Mount Thera in the Aegean Sea, which is known today to have occurred at about 1650 B.C. The reason for connecting these two events is the claim that the volcanic explosion caused a tsunami (a flood by sea waves caused by a submarine earthquake) which drowned the Egyptian army. Lately, ashes from this eruption were found in the clays of the Egyptian Delta and this finding was also related to these two events. The date of 1650 B.C. was derived from examination of cores taken from the glaciers of.Greenland and dated by 14C (Hammer et al. 1980, Dansgaard 1985). The ice shows traces of acidity caused by this volcanic explosion. Although it is apparent that the explosion of Mount Thera had an environmental effect on the eastern Mediterranean coasts, either as flooding by waves or as ash falls, yet the author does not suggest fixing exact dates, but examining the possibility that there were anomalous environmental and historical events, which remained in the collective memory of the people, later to be connected together, in a cause and effect bond.

It is interesting to note that the precipitation curve from Soreq Cave the levels of the Dead Sea and the humidity of Lake Van show a rise in the precipitation at this period of time. This is parallel with a certain advance of the glaciers of Scandinavia (see Figs. 1.2 and 1.5)

Thus in the first place all these observation are an additional support to conclusion brought up, time and again, in the present book, namely that the biblical stories have a core of truth, which is based on historical events. These events left their traces on the environment and thus may help in exposing the core. In the case of Exodus the chain of events seems to be the following:

A cold spell, presumably but not necessarily, was caused by the volcanic explosion. If this was the reason, the chilling of the globe was due to the fine ash being dispersed in the atmosphere and reducing the solar radiation reaching the earth.

This caused the glaciers to expand, which caused the movement southward of the belt of the Westerlies. This may have pushed southward the inter-tropical arid belt, causing a reduction in the precipitation over Ethiopia and thus less water to flow in the Nile, which also meant less water and food to spare for any alien population.

Further evidence of a cold spell is from research carried out in the Libyan Desert by a German team from the University of Berlin. It shows that, at about 1420 (+ 230) years B.C. (dated by Carbon 14), a humid spell affected this desert. On the other hand, more or less at that time (between 1500 and 2000 years B.C.), a dry spell affected Kenya (Pachur and Roper 1984).

The southward movement of the arid belt and, in its wake, the belt of the northern low pressure, may have had another effect on the natural environment of Egypt, like that which affected it during the Last Glacial Period. Rain storms coming over the Mediterranean would penetrate southward into the deserts of Cyrenaica and Egypt causing rain to fall on areas which for centuries had received only a few millimeters of rain. Such rain would cause severe changes in the natural habitat, as well as the land forms. The dates of the climatic changes discussed earlier were fixed by radiocarbon methods in which an inaccuracy of about 100 years is reasonable. This means that if Exodus was really connected with this climatic change, it cannot be placed exactly, but sometime at the beginning of the 18th Dynasty, which started 1575 B.C. In the Bible we read about ***A new king over Egypt, which knew not Joseph.*** Exodus: 1: 8). The other question is how long did the enslavement and strife last and when did the Hebrews begin to leave Egypt? We do not know for sure whether they left in one group, as is told in the Bible, or by several emigration waves. If the period of hardship lasted 200 years, Exodus could have happened, for example, at the time of Akhenaten (1362–1345 B.C.), the king who declared the sun-Aten as the chief god over Egypt. It could have taken place during the reign of Tutankhamun (1344–1335 B.C.), the well-known Pharaoh whose tomb was found untouched about a century ago. These were weak kings who, apparently, were not able to organize a campaign outside Egypt in order to force the Israelis back into slavery.

The suggestion of the author is thus to connect Exodus with a humid period after the climate change which took place after the dry Late Bronze Period. This period of humid climate, which marked the start of the Iron Age started about 1300 B.C., as can be deducted from the paleo-climate curve from the Soreq Cave and the rise of the level of the Dead Sea (Bar-Matthewes et al. 1997).

The environmental impact on Egypt, the core of the story of the 'Ten Plagues' is the following:

The heavy rains over northern Egypt cause foods in the wadis or ephemeral river-beds of this part of Egypt, during the winter seasons. These wadis are mostly dry during normal arid periods if they run it is for short periods and not simultaneously. These floods which occurred in abundance eroded and carried silts and soils from the surrounding hills into the clear water of the Nile, causing it to become brownish, or even reddish. This took place during the winter season in which the Nile is low and relatively clear. As mentioned, the local name for the soils of northern Egypt was "Deshret", namely "Red". The inflow of "Deshret" soil, will color the water and in the same time, would make it undrinkable. Moreover, the water heavily laden with silts and soils would kill the fish in the river and in the ponds and lakes adjacent to it. ***And all the waters that were in the river turned to blood. And the fish of the river and there was blood throughout the Land of Egypt*** (Exodus: 7: 20–21).

The humidification of the land due to heavy rains would have caused an increase in the bio-world. The habitat of frogs would have expanded beyond the banks of the Nile. ***And the river shall bring forth frogs abundantly, which shall go up and come into thine house, and into thy bedchamber*** (.Exodus: 8:3).

The author and his family had a similar experience when in 1975 they moved from Jerusalem to live at Sede Boqer Campus, at the Institute for Desert Research of the Ben-Gurion University. This campus is located in the middle of the Negev desert. The introduction of the new system of drip-irrigation, which caused the almost permanent humidity of the soil, brought also an explosion in the population of toads.

The humidification of the extremely dry climate of Egypt would also have caused an increase in the number of parasites, such as lice and bacteria causing plagues infesting man and his animals. And… ***the dust of the land… become lice throughout all the Land of Egypt*** (Exodus: 8:16) … ***and there came a grievous swarm of flies… the land was corrupted by reason of the swarm of flies*** (Exodus: 8: 24). ***The hand of the Lord is upon thy cattle which is in the field, upon the horses… there shall be a very grievous murrain*** (Exodus: 9: 3). ***… and shall be a boil breaking forth with blains upon man, and upon beast***. (Exodus: 9: 9).

The storms, coming as rain storms, push heavy clouds of dust ahead of them, darkening the light of day, and are followed by heavy hail, lightning, and a heavy load of static electricity in the air. This would be a very uneasy experience even for a geologist from a more humid region, how much more so for a people who were born and grew up in a land in which the annual precipitation is almost nil… ***and the Lord sent thunder and hail, and the fire ran along upon the ground and the Lord rained hail upon the Land of Egypt*** (Exodus: 9: 23). ***And Moses stretched forth his hand toward heaven and there was a thick darkness in all the Land of Egypt…*** (Exodus: 10: 22).

The author and his team experienced such darkness in the middle of the day, while travelling along the eastern shore of the Suez Gulf, when a dust storm coming from Egypt enveloped their line of the vehicles. The people in the cars lost vision not only of the other cars in the line but the dust was so heavy that those sitting in the back sits of the vans could hardly see their fellows (Fig. 12.2).

Describing the story of the travels of the patriarchs and that of Exodus from a paleo-environmental point of view, the story would run as follows:

The decline of rainfall over the countries of the Western Mediterranean at about 2000 B.C. caused the Hebrews to immigrate southward into Canaan, on their way to Egypt. At about 1750 B.C., the Hebrews joined the Hyksos movement into Egypt in order to escape another phase of drought that caused famine throughout the Fertile Crescent.

Fig. 12. 2 A dust storm followed by a violent rain storm over the Eastern Mediterranean, covering northern Egypt, December 2010. (Courtesy NASA)

At about 1300 B.C. the climate began to change. Canaan became more humid due to the movement of the climatic belts southward, it caused the sub-tropicl drainage basin of the Nile to become dryer, which brought a fall in the level of the Nile. This caused famine in Egypt. At the same time it caused dust/rain storms to penetrate into northern Egypt destroying crops, eroding and sweeping red soils, from the surrounding hills, into the Nile. Humidity caused propagation of parasites causing plagues. This gave the Hebrews the opportunity to escape serfdom, as the desert of Sinai, which separated their land of enslavement from the land from which their ancestors came, turned passable. The ample and more frequent rains over this desert held the promise of an increased water supply and better grazing.

References

Bar-Matthews M, Ayalon A et al (1997) Late Quaternary Paleoclimate in the Eeastern Mediterranean region from stable isotope analysis of Speleothems at Soreq Cave, Israel. Quatern Res 47(2):155–168

Dansgaard W (1985) Greenland Ice Core Studies. Paleogeog Paleoclimatol Paleoecol 50:185–187

Hammer CU, Clausen HB et al (1980) Greenland Ice Sheet Evidence of Post-Glacial Volcanism and its Climatic Impact. Nature 288:230–235

Pachur HJ, Roper HP (1984) The Libyan (western) Desert and Northern Sudan; during the late Pleistocene and Holocene. Berliner Geowiss. A.50

Wilson I (1985) The Exodus enigma. Weidenfeld and Nicholson, London

Chapter 13
Crossing the Land of Swamps and Reeds

Abstract In order to avoid being harassed by the Egyptian army, the Hebrews did not take the shorter route along the coastline, as this road was lined with forts and Egyptian war-chariots were patrolling it. They took a more southern route and encamped along the western border of the area of swamps and reeds. A fierce dust storm took place, which dried up the gypsum and salt crust of the soil surrounding the swamps, and which enabled the Hebrews to cross them. When the Egyptian chariots moved after them their wheels broke the crust and they sank into the marshes.

> And it came to pass, when Pharaoh had let the people go, that God led them not through the way of the land of the Philistines, although that was near for God said, Lest peradventure the people repent when they see war, and they return to Egypt: But God led the people about, through the way of the wilderness of the Red sea: and the children of Israel went up harnessed out of the land of Egypt (Exodus: 13: 17–18)

After the story of the "Ten Plagues" come the stories of the crossing of the Sea of Reeds (*Yam-Suf* in Hebrew) or Red Sea and that of the wanderings of the Children of Israel in the desert of Sinai. Although there is no intention of drawing a map of the route the tribes took through the desert, still the general characteristic of the area seem to fit well with the philosophy of the present book to understand the natural environments through which the biblical stories flow and the impact of climate changes on these environments and thus on the people. In this chapter the details of the biblical narrative of the events following the Exodus will be examined from the environmental and climatic points of view.

In the first place it has to be clarified that the term "sea" in the Bible refers to a body of standing, as opposed to flowing water. Lakes and even large water tanks, like the one, which King Solomon set up at the entrance of the Temple in Jerusalem, were called "sea". Thus also the "sea" crossed by the tribes of Israel on their way from Egypt to Sinai, which is called *Yam Suf'*, namely the Sea of Reeds, could have been a lake or even a swamp along the shores of which grew reeds which need a source of fresh or brackish, but not saline sea water. It can be assumed, thus, that the Biblical *Yam-Suf* or Sea of Reeds was the area where the

A. S. Issar, *Strike the Rock and There Shall Come Water*, SpringerBriefs in Geography, DOI: 10.1007/978-3-319-01937-6_13, © The Author(s) 2014

Suez Canal passes today. Before the canal was dug this land strip, occupying the rift valley between the Red Sea and the Mediterranean, was an area of swamps, In Arabic a swamp is '*sabkha*', while a sea and a lake are '*bakhakr*', or 'bah'r', of which the Gt. Bitter Lake still exists (Fig. 12.1).

Towards the end of the wandering, when the tribes reach the eastern side of Sinai, they come again to *Yam Suf.* This time it is the eastern branch of the Red Sea, which continues northward as the Arava Rift Valley. Also along this valley existed and still exist *sabkhas*, formed by groundwater emerging from deeply buried aquifers to reach the surface of the land. These deep aquifers, contain brackish water under artesian pressure and were discovered by oil and water exploration wells drilled in this region. It is thus fairly safe to assume that the swamps between Egypt and Sinai, i.e. the Biblical Sea of Reeds erroneously translated to Red Sea emerged also in the days of Exodus. Today such an emergence of groundwater occurs at the southern edge of this area, in the form of the springs of Uyun Musa, namely the "Springs of Moses". These springs emerge along the central and northern part of the fault, forming swamps surrounded by reeds. Moreover, the sands covering the area east of the rift valley also contain a shallow water table of brackish water which could have risen during a spell of humid years to form marshes, and even lakes surrounded by reeds. Such lakes are found today during a very humid winter, in the valleys between the sand dunes east of the Suez Canal. Thus, it seems very probable that a more humid period would create an environment of marshes and lakes in the region separating Egypt from Sinai which is the border between Africa from Asia.

It should not be forgotten, however, that the term humidity in a desert is a very relative term. Even during a humid spell this region was still arid; precipitation, even if double that of today, would not exceed 100 mm. In such an arid environment, *sabkhas* or swamps are characterized by the formation of a crust of evaporates such as calcium carbonates, gypsum, or halite (common salt).

According to the chemical analysis of the water found in the deep aquifers and springs, the ancient *sabkhas* along the sea shore were probably rich in calcium sulfate (gypsum). Such deposits are found today in this region around the springs of Uyun Musa and more to the south around the springs of Saidna Musa ('Our Master Moses'). These springs emerge along the faults bordering the Suez rift on its eastern side, forming *sabkhas* with an evaporitic crust of gypsum. Another chemical component of this water is magnesium sulfate which gives it a bitter taste.

With this hydro-geological and hydro-chemical background, the plight of the Tribes of Israel, during their passage through this region can be better understood. The Bible tells us about a large group of people coming from the fertile Land of Egypt trying to escape the Egyptian army. The season is spring which is typified, even today, by rainstorms following hot, dry spells or *khamsin* which is a dry wind heavily laden with dust, blowing from the east.

We are told that the tribes escaping from Egypt preferred not to take the northern, shorter route along the coastline, as this was the Road of the Kings or the Land of the Philistines. (As mentioned above, another example for the later editing of the verbal stories by the scribes, as the Philistines did not arrive until 1200 B.C., namely during the reign of the 20th Dynasty). The reason given by the Bible is supported by archaeological evidence, which shows that this road was lined with and patrolled by Egyptian army garrisons. It was called by the Egyptians: The Route of Horus. Indeed, archaeological investigations as well as Egyptian documents tell about a line of forts and war-chariots patrolling along this road.

To avoid meeting Pharaoh's war chariots the Hebrews took a more southern route which brought them into an area infested with swamps, salt marshes, *sabkhas*, and thickets of reeds. Reaching this area, they heard the report that the Pharaoh's chariots they had tried to avoid were following them, to march them back into slavery. They encamped along the western border of these stretches of swamps and marshes. At that time a fierce dust storm, typical for this time of the year (but probably even more fierce than in our time, due to the different climate) blew from the east, reducing the visibility to nil and stopping the Egyptian army from overtaking them as ***the pillar of cloud went from before their face and stood behind them*** (Exodus: 14:19). The same hot, dry storm dried up the gypsum and salt crust of the soil surrounding the swamp, and the sea goes back ***by a strong east wind all that night and made the sea dry land and the waters were divided*** (Exodus: 14: 21) most probably by a hardened crust that formed by a layer of gypsum. This might have given the Israelites an opportunity to escape the Egyptian army by going westward into the maze of marshes. When the Egyptian army moved after them it could not envelope them from the flanks, as any cavalry or chariot army would have preferred doing to a swarm of pedestrians, because of the marshes ***which were a wall unto them on their right hand and on their left*** (Exodus: 14:22). In this context, a wall would mean a protection on their flanks. The chariots could only move behind them and try to overtake them on the dried-up tracks. ***And the Egyptians pursued and went in after them to the midst of the sea, even all Pharaoh's horses, his chariots, and his horsemen*** (Exodus: 14:23).

At this point two phenomena might have taken place. First, the crust which would bear the weight of man could not carry that of the chariots, the wheels of which started to sink into the crust (an experience the author has had two or three times in *sabkhas* of this type, until he learnt the lesson that where a geologist can walk, a jeep may sink up to its shaft). This sinking delayed the Egyptian army in the midst of the *sabkha*. ***It took off their chariot wheels that they drave them heavily*** (Exodus: 14:25). At that time the *khamsin* may have "broken", namely, the dry, hot eastern wind was the precursor of a low pressure rainstorm coming from the sea and flooding the area. ***Thus the sea returned to its strength when the morning appeared and the Egyptians fled against it*** (Exodus: 14:27).

This scene could have happened along any part of the valley between the present cities of Suez and Port Said, crossed today by the Suez Canal, as all this area was most probably infested by such swamps. A rise in the groundwater table in the water-bearing layers adjacent to the Suez Rift Valley is also a possibility. This caused the area to become even more swampy with many springs of brackish water so that reeds grew around and in the midst of the swamps. Furthermore, the regime of heavy dust- and rainstorms following one after the other was much more intensive than at present. If one adds to this the relatively high gypsum content of the water emerging from the aquifer and brought by the floods, and the drying effect of an eastern dust-laden wind ushering in such a rainstorm, the stage is set for a scenario such as that described in the Book of Exodus. As previously mentioned, one should remember that a regression of the sea due to water caught in the glaciers which spread in the polar regions during that period could also have changed the geography and environment of the area between Egypt and Sinai as well as all along the shores of Egypt. A regression of the sea in the order of magnitude of a few meters would have changed all the travel alternatives between Egypt and Sinai.

This is another reason why the author is reluctant to draw a map indicating the routes along which, the Israelites wandered in Sinai. The first reason has already been stated, namely, that the author does not suggest taking the stories of the Bible as scientific reports, but they are only an echo of core story of the traumatic experience the people went through. Moreover, it is claimed that the fiercer the experience, the more it will portray the environmental background. (The reader who is interested in the routes of Exodus is advised to consult Harel (1976) and Wilson (1985).

Assuming a change of climate did take place, another question which comes to mind is how it influenced the Desert of Sinai? From the Bible story sounds the echo of a journey through a formidable desert and the suffering of the wanderers from lack of water. How do these echoes agree with the assumption of a change in climate? The answer to this lies in understanding the nature of such a change and thus its impact on the desert. It is assumed the desert would not have changed into a humid place, but would have produced a special type of climate controlled by a general southward movement of the arid belt. Yet, during summer this belt moves northward. Thus, the region still maintains its arid character of long, hot summers which might be interrupted by a few rainstorms. During autumn, winter, and spring, frequent spells of rainstorms following heavy dust storms would occur. This is due to the seasonal movement of the arid belt southward. From the ecological, environmental point of view, such a climate might affect the quantity and frequency of the floods between autumn and spring, increase the amount of water flowing from the springs, and increase the vegetation cover, especially in river-beds and in the mountain regions. Thus, more grazing and firewood can be found in the desert. It would not, however, make the desert a fertile land or more pleasant, especially during the summer months. The author is of the opinion that a picture of this type of desert is drawn in the story of the wanderings of the Tribes of Israel in the deserts of Sinai and the Negev.

References

Harel M (1976) The Exodus route in the light of historical-geographic research. In Geography in Israel, 23rd Int. Geog. Cong. USSR

Wilson I (1985) The Exodus Enigma. Weidenfeld and Nicholson, London

Chapter 14
That Great and Terrible Wilderness

Abstract The route of the tribes from Egypt to Canaan, through central Sinai Desert was decided by the location of sources of water. Taking into account that the climate was more humid than the present, then during most of the year the springs would flow abundantly and enough forage could be found. Problems of supply would have to be faced, either when a big concentration of people stays in one place or towards the end of the summer, when small springs may fail and vegetation diminishes. In such cases, the people will try and find springs and fodder in the higher mountains where more precipitation promised more water and food. Another problem to be overcome is that described in the case of Marah, the first water source encountered and where brackish water emerges which was not to the liking of the people. These springs and others more to the south are not fed from local rains or floods but from deep fossil water bearing strata.

> So Moses brought Israel from the Sea of Reeds and they went out into the wilderness….and found no water. And when they came to Marah they could not drink the waters, for they were bitter therefore the name of the place was called Marah (Exodus 15: 22–23)

In the desert, one is not surprised when he reaches a spring and finds the water salty. This is due to the fact that the rain water, brought by storms driven by winds coming from over the sea contains some salt. When this water evaporates it leaves the salts on the surface. The water from the following heavier rain storms will dissolve the salts and infiltrate to enrich the ground water. Part of the water on the way down will be caught by the roots of desert shrubs, which will by osmosis leave part of the salts in the sub-soil. These will be flushed down to the ground water by the infiltrating water of a very heavy storm which occurs from time to time, or by water of a flood due to such a storm. Thus, it is interesting to note that the Bible speaks about the taste of bitterness, which the Children of Israel complained about and not the saltiness. This is evidence that the story of Exodus contains nuclei of experiences of the Children of Israel after leaving Egypt. The bitter

A. S. Issar, *Strike the Rock and There Shall Come Water*, SpringerBriefs in Geography,
DOI: 10.1007/978-3-319-01937-6_14, © The Author(s) 2014

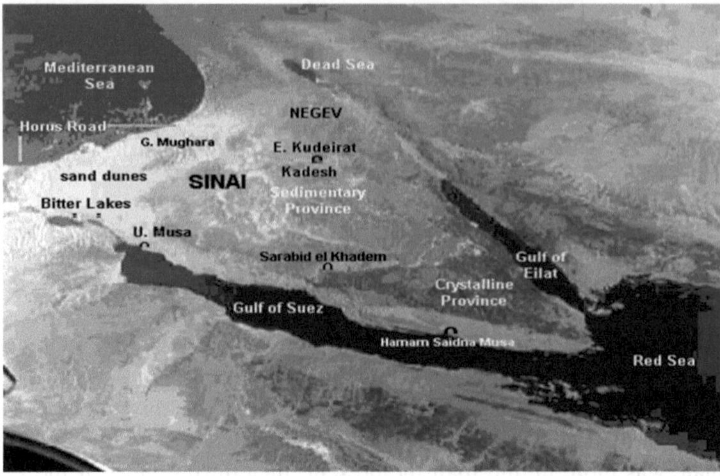

Fig. 14.1 Sinai and negev deserts as seen from space (Courtesy NASA)

taste of water in arid regions is, in most cases, a function of the relative abundance of sulfate salts, especially magnesium sulfate ($MgSO_4$) (Epsom or English salt) which, even as a small constituent of the salts, will give the water a bitter taste. As already mentioned, indeed the water of the large springs of Uyun Musa (Fig. 14.1) which can supply drinking water for a multitude of people and which are located on the northeastern edge of the Red Sea is rich in sulfate (500 ppm) mainly calcium sulfate, but also contains magnesium sulfate which gives a bitter taste.

Thus this story can be marked as one of the first, if not the first, "hydro-chemical" reports. If one accepts the identification by travelers and geographers of Uyun Musa with Biblical Marah and continues to investigate the secret behind the emergence of these large bitterish springs on the border of the desert, one will arrive at a very interesting hydro-geological story which began about two and a half million years ago and still plays an important role in the environment of these deserts.

The best way to understand the impact of the hydrology on the environment is to view the region from space, giving a good picture of the aridity of the deserts of Sinai and the Negev (Fig. 14.1).

A more detailed image shows the site of the springs of Uyun Musa as dark specks due to the vegetation cover (Figs. 14.2 and 14.3). In deserts, a concentration of vegetation denotes places where groundwater emerges.

In the previous chapter, the emergence of groundwater along the rift valleys of the Suez and Arava was mentioned. The following is a description and explanation of these phenomena in more detail and thus to understand better the impact of the availability or lack of water resources on the fate and faith of the tribes traveling through these areas.

Looking at these satellite images of Sinai and Negev one's attention is drawn to the triangular shape of the Sinai Peninsula and the Negev, which together form

Fig. 14.2 The oasis of uyun musa (photo by A.S. Issar)

Fig. 14.3 One of the springs of uyun musa (photo by A.S. Issar)

a transition zone between Asia and Africa. One can observe that the two regions, divided by the political border between Egypt and Israel are not divided by a geographic boundary, either a valley or ridge. In fact, they form one geographical unit, from both a geological and a topographical point of view. Also, both areas have features in common with the regions bordering them to the north and the west: the Sinai Peninsula is similar to the eastern and western deserts of Egypt and the Negev is similar to southern Israel, Jordan and northern Arabia.

This resemblance signifies the special character of the Sinai Peninsula and the Negev. These regions existed as one geological unit during most of their geological history and were divided from Asia on the east and Africa on the west only comparatively recently. The division was effected by down breaking of the Syrian-African rift system, with the rift containing the Gulf of Suez on its western edge and the Gulf of Eilat and the Dead Sea rift on the eastern edge. This separation of Sinai and the Negev from Africa and Arabia occurred at the dawn of the Quaternary era, some two and a half million years ago, rather a short while ago in geological terms. These deep troughs, or rift valleys, formed by the shearing and the down-faulting movements, continuing through the Lower and Middle Quaternary, gave the rift valleys their present morphological features and caused the diversion of the drainage patterns from the Mediterranean Sea to the rifts (Deut. 1:19, Garfunkel and Bartov 1977; Garfunkel and Ben Avraham 1996). This rifting also profoundly affected the subsurface flow of groundwater. The regional faults along the rift valleys' borders fractured the rocks along the faults and vertically displaced impermeable layers, thus opening flow routes between the deep subsurface and the surface. Water still uses these routes to flow from the depths to the surface to form springs (Issar et al. 1971).

Due to their in-between character, the Sinai and the Negev maintain geologic, structural, and stratigraphic affinities with the African and Arabian plates, and can be regarded as a miniature plate which exhibits, on a small scale, many features of the larger plates. On the southern edge of the Negev–Sinai Peninsula, one finds a triangle of the basement shield, built mainly of crystalline rocks of Pre-Cambrian age (Fig. 14.1). This part of the Sinai is mountainous. Mount Sinai is traditionally located in this region. Northward, an area built of sandstones overlies the crystalline basement. These sandstones were deposited after the lower Paleozoic or mid-Mesozoic on a wide continental stable zone, seldom invaded by the sea. When sea intrusion did occur, thin shallow marine deposits were left behind.

The highly permeable sandstone rocks enable the infiltration of water, and at depth form a series of water-bearing strata. The belt of sandstone layers is bordered to the north by a steep escarpment formed by limestones of Middle Cretaceous age. These are overlain by chalks of Upper Cretaceous age and limestones and chalks of Tertiary age. The plains of central Sinai are rolling hills with occasional mountains protruding from the plain. The hills are built from nearly horizontal layers of chalks and marls of Upper Cretaceous to Tertiary age, while the mountains are anticlines of folded limestones of Middle Cretaceous age. In some of these anticlines, the cores have been eroded and sandstones and limestones of Lower Cretaceous to Jurassic age are exposed. The chalks and marls are impermeable to water, and thus springs are scarce in the plains of central Sinai. The valleys between the hills are filled mostly by desert loess (an air borne silt deposited during Upper Quaternary times) and some gravel and sand layers. The northern-most belt of the Sinai Peninsula and its extension into the Negev is formed from a plain covered by sand dunes (Fig. 14.1).

The scanty rains (50–100 mm per year) falling on these dunes infiltrate quickly into the sands and thus do not evaporate. This water accumulates on the impermeable layers underlying the sand dunes and thus forms a shallow groundwater table with a gentle gradient toward the sea. Near the coast, the groundwater table is very close to the surface of the land, and where it outcrops, or where the inhabitants dig shallow wells to reach this water, oases are found. These oases determined the routes taken by people crossing the deserts. The ancient routes connecting Egypt to the Fertile Crescent of the Middle East were along the coast of the Mediterranean where there are strings of oases fed by the shallow perched groundwater table. El-Arish is the main oasis on this desert crossing. The groundwater here is also fed by the occasional flooding of Wadi El-Arish, which drains most of central Sinai. The water spreads over the flood plain of the wadi, percolates to the subsurface gravel beds, and from there to the sandstones and sands of Quaternary age.

A well-guided traveler in ancient times could depend on a reliable supply of water for his party and his pack animals over most of the northern route. He would have to carry only enough water to help him cross some of the longer stretches between the oases. However, crossing the Sinai desert becomes more difficult for even small groups of travelers, not to mention large groups, if they must choose a more southern course. Water resources are scarcer along this route than in the north. Only two clusters of large springs, sufficient to supply a multitude of people, are to be found. One cluster is that of Uyun Musa springs near the Gulf of Suez (Fig. 14.1). The other is the cluster is the Ein Qudeirat ein Qadis Uyun el Quseima springs (apparently the Biblical Kadesh Barnea) near the border of Israel (Fig. 10.1).

Imagining the route of the Israelites from Egypt to Canaan, through central and southern Sinai avoiding the road along the coast but still assuring sources of water for the wanderers, will first of all include the group of springs of Uyun Musa, discussed already. What is the secret behind this rich flow of water at the edge of the desert? A short survey will show that these springs are not fed from local rains or floods; their water's source is deep water-bearing strata which stores a huge quantity of water in its pores. The water would be pushed up to the surface by artesian pressure. A change in this pressure may be effected by a change of climate, namely, by more rains falling on the outcrops of the aquifers in southern Sinai and the water infiltrating the subsurface, raising the water table (Nativ et al. 1987).

Such artesian aquifers are found under many of the deserts of the world, such as the Sahara Ambroggi 1966; Klitzsch et al. 1976; Issar 1985). In many instances their outlets are along fault lines. This water can be defined as "fossil-" or "paleowater" since it is many thousands of years old. The age was determined by the isotope of Carbon 14 content of the carbonate dissolved in the water. In most cases the age is in agreement with the time-span calculated on the basis of hydrologic flow models. Such a time period would have been necessary for water which had infiltrated the subsurface in the outcrop areas to flow to the desert oases, located hundreds or even thousands of kilometers away. Yet pressure waves will travel faster. These will affect the flow of the springs thousands of years before the recharged water will reach the outlet. One can imagine the difference between the velocity of the flow of water and the velocity of the pressure wave which affects the flow of the springs as the difference between a sound wave and the velocity of the wind.

It is no wonder that Moses led the tribes toward these springs after entering Sinai. He knew from his previous wanderings in the desert that there was an ample supply of water. At Uyun Musa, we are told, however, that the people were sorely disappointed as the water tasted bitter, especially after the fresh Nile water to which they had been accustomed to all their lives. We do not know how Moses managed to sweeten the water. Based on personal experience, the author knows that, as the thirst grows, one loses one's sensitivity to the bitterness of the water, and as one continues to drink the water for a long while, one loses even the sensitivity to the bad effects of the magnesium sulfate in the water.

After becoming acquainted with the landscape of the Desert of Sinai and its geological and hydrological features, one can imagine the problems facing any leader or commander who must plan the crossing of this desert by a multitude of people. The first problem, undoubtedly, would be the supply of water and food for the people and their animals.

Taking into account the possibility that the climate was more humid, with a precipitation perhaps three times than today, this would raise the quantity of rain from 50 to 150 mm. One can assume that during winter, spring, and early summer, most of the springs would flow abundantly and enough forage could be found on the desert plains. Problems of supply would have to be faced, either when a big concentration of people stays in one place or towards the end of the summer, when small springs may fail and the desert grasses are depleted. Problems of water supply may occur on the long stretches between one group of springs and another, or when a year or two of drought may deplete the flow from the springs with a small storage in their subsurface. In such cases, the leader would have to disperse his people to try and find springs and fodder in the higher mountains where more precipitation, and even snow in winter, promised more water and food. Another problem to be overcome is that described in Marah, namely, the bitter taste of water caused by the sulfates. However, such springs are encountered only along the rift valley, namely, at the beginning of the wanderings and might also have been encountered toward the end before crossing into Moab.

As there is no intention in the present book to present another Bible exegesis, as enough books in this category were already written, no attempt will be made to draw the "true" route of the Israelites through the Sinai. Nor is there an intention to explain the miracles, which occurred to the Israelites. The main intention is to reveal the cores of events which are described in the Bible and can be understood by the scientists who investigated the water resources of Sinai, and later the impact of climate changes during history on these resources all over the Middle East.

Thus, analyzing Sinai from the hydro-geological and environmental point of view, one would say that only a few places exist suitable for the concentration of many people. The first place is around Uyun Musa, where the brackish springs emerge. Another is between the sand dunes northeast of Uyun Musa, A third is in the vicinity of the present small town of A-Tur where a warm brackish spring emerges, the name of which is Hamam Saidna Musa, (The bath of our master Moses) (Fig. 14.1). Here, many palm groves flourish because of the shallow freshwater table that exists in the vicinity of these springs. Also, in river-beds between Uyun Musa and Hamam Saidna Musa clusters of small springs are found, supplying sufficient water for small groups of people, but not for a multitude over a long period.

An area which might have been rich in springs is the erosion cirque at Gebel Muaghara where rocks of Nubian Sandstone are exposed and where a shallow water table also exists. Another place would be in the south, namely in the region built of crystalline rocks, such as granites, gabbros, and porphyry (Fig. 14.1).

In this region there are many small springs and oases, the biggest of which is Wadi Feiran where many groves of palm trees grow.

As mentioned, after crossing part of Sinai, another group of perennial springs and a large area with shallow groundwater where the tribes camped at 'Kadesh Barnea', believed to be the oasis of Ein Qudeirat, at present the site of the small town of El-Quseima. This place is described as the main base of the tribes of Israel, before their first attempt to invade Canaan via the Negev.

According to the Bible, after leaving the springs of Marah and camping at several watering places, the tribes of Israel reached Mount Sinai, termed also Horeb, where they had a big gathering. At this site Moses went up the mountain for the Tablets, the receipt of which was connected with thunder, lightning, fire, and the loud call of a trumpet.

As a scientist who worked, for quite a few years in the deserts of Negev and Sinai such phenomena are not regarded as un-natural. During spring and winter, rain-storms over the northern and central part of the country, reach the desert as heavy clouds and as storms of lightning and thunder, which seldom terminate in a heavy shower or even hail. If one happens to be on a high mountain, when such a storm reaches the desert the unforgettable experience is not pleasant. Such a storm of lightning, which was experienced by the author, caused the atmosphere to be loaded by a static electrical charge which caused the hair of the members of his team to stand and when raised, their geological hammers buzzed.

Another natural phenomenon which may have impressed the Israelites in the Sinai Desert is an earth-quake which occurs there from time, especially along the regions bordering the Red Sea, as these areas are built of tilted blocks of sedimentary rocks which contain oil deposits, some of which seep up to the surface. It is quite possible that during the time the tribes were in this region an earthquake could have caused the emergence of oil and gas which caught fire and blazed. For centuries, such phenomena were known to occur in western Persia and were connected with fire worship. Thus, a combination of a few dramatic natural phenomena occurring while the tribes were in the Sinai Desert could explain the stories narrated from father to son and later written down in the books of Genesis and Exodus.

The location of Mount Sinai is a subject upon which there is no agreement between scholars. At present there are a few high mountains in the Sinai and even the Negev, which, based on scholarly arguments, are claimed to be the mountain on which the Ten Commandments were given, as for example Professor Menashe Harel's suggestion (1981) that it is *Mount Sinn Bisar* in northwestern Sinai. He bases this location on its distance from the crossing of the Sea of Reeds, and the distance women, children and herds can go in the three months after Exodus (Exodus 19:1). In the opinion of the author, the story of the Revelation on Mount Sinai is an agglomerate of a few events which occurred to a few of the tribes, on their way through Sinai. The core of the story might have been either an electric-atmospheric anomalous event like the one, which the author experienced and described above, and which took place near or on one of the mountains along the route of the Israelites. Another possibility is an earthquake along one of the regional fault lines criss-crossing the Peninsula. This anomaly or a chain of anomalies was later connected with an attempt to revive a certain pagan cult brought from Egypt and followed by an epidemic, which happened due to the pollution of the water or the food. All these happenings were interwoven when told from father to son, into one awesome marvelous event to be remembered for generations to come.

The heart of the story is the ascent of Moses to the top of the mountain, and his disappearance for forty days and nights. While awaiting Moses descent from the mountain the people felt abandoned in the center of the desert and they asked his brother Aharon to: ***…make us gods, which shall go before us for as for this Moses, the man that brought us up out of the land of Egypt, we wot not what is become of him*** (Exodus 32:1). This led to the making of the Golden Calf, which was accepted by the people enthusiastically, most probably because it was a known and even admired god or goddess. The reason could be connected with the Egyptian cow-goddess Hathor. The center of Hathor worship was in the turquoise mines of Sarabit-el-Khadm where Israelites were working as slaves.

References

Ambroggi R (1966) The Water under the Sahara. Sci Am 214:21–29

Garfunkel Z, Bartov Y (1977) The Tectonics of the suez Rift. Geolog Surv Israel 71:44

Garfunkel Z, Ben-Avraham Z (1996) The structure of the Dead Sea basin. Tectonophys 266:155–176

Harel M (1981) The Sinai journeys: the route of the Exodus. Ridgefield Publisher, San Diego

Issar A, Rosenthal E, Eckstein Y, Bogoch R (1971) Formation waters, hot springs and mineralization phenomena along the eastern shore of the Gulf of Suez. Bull Int Assoc Sci Hydrol XVI (3):25–44

Issar A (1985) Fossil water under the Sinai Negev Peninsula. Sci Am 253 (1): 104–112

Klitzsch E, Sonntag C et al. (1976) Grundwasser der Zentral Sahara Fossil Vorrate. Geol. Rund., Stuttgart 65:264–287

Nativ R, Bachmat Y et al (1987) Potential use of the deep aquifers in the Negev Desert, Israel—a conceptual model. J Hydrol 94:237–265

Chapter 15
To Smite the Rock and Water Shall Come Out

Abstract The groundwater flow in the crystalline rocks of Sinai explains the story of striking the rock to obtain water. It looked miraculous to people coming from a land where the only source of water was the Nile and shallow shafts dug nearby. Kadesh Barnea probably the present site of the spring of Ein Kudeirat, was the site where Miriam the sister of Moses passed away, and where the other incident of striking the rock took place. From this site the mission went to spy out the Promised Land and the first failed attempt to storm the land, was made. This brought the Israelites to avoid a direct assault against Canaan and made a detour through Transjordan. Avoiding war with the Edom, Moab and Ammon, conquered the Gilead and Bashan where a few tribes settled.

> Behold, I will stand before thee there upon the rock in Horeb; and thou shalt smite the rock, and there shall come water out of it, that the people may drink. And Moses did so in the sight of the elders of Israel (Exodus.7:6)

Many times when the author saw the Bedouins of southern Sinai excavate their wells in the crystalline rocks, from which this part of the peninsula is built, the story of Moses striking the rock to get water came to mind. These Bedouins, belonging to the tribe of the Gebelia (mountain people), live around the monastery of Santa Katerina and, according to their tradition, which has been confirmed by historical research, were once Christians who were brought by the Byzantine emperor Justinian, from the Balkans in the sixth century A.D. to be servants to the priests of the monastery. They know how to discern places where veins of calcite filled the fractures of the granites such places are a sign of an extinct spring. They also know how to distinguish an acid hard granite rock, and hard porphyry dike from a soft diabase dike. The brittleness of the rock enabled the Bedouins to dig along the wall of granite bordering the dike.

The moral of this story is that when the Sheik of the Gebelia Abu-Heib was asked to explain how water is found in a rock, his answer was "min Allah" which, literally translated, means "from God". Apparently he did not bother to delve into the problem of the flow of water through fractured media, not because it is much more complicated than that of flow in porous media, but because, for him, as for his

A. S. Issar, *Strike the Rock and There Shall Come Water*, SpringerBriefs in Geography,
DOI: 10.1007/978-3-319-01937-6_15, © The Author(s) 2014

ancestors and the ancestors of the author, all phenomena, especially those connected with water in the desert, involved the direct intervention of a divine power.

The author received a similar answer when he asked the mullah of an Iranian village in the mountains of Kashan to explain the pulsations of the karst spring irrigating the orchards of the village. A Zoroastrian shrine, still standing above the cave from which the spring emerges, was evidence that the Persian ancestors of the mullah and his congregation also believed that a divine power, and not merely a hydraulic siphon, was involved in causing the spring to pulsate. Whether this power came directly from Ahura Mazda, or whether the responsibility was bestowed on a minor divine power, needs further investigation. In the western countries of the Fertile Crescent, the pulsations are attributed to the presence of a gin or genie (a goblin or little devil) inside the spring. Such a spring is the Gihon spring, which was Jerusalem's main source of water and, as will be told later, the water that the Canaanites, and later the Jews of Jerusalem used for libation.

The repeated stories of the wanderers' sufferings from thirst in the desert are not surprising albeit the various evidence that the period of exodus was a relatively humid period. As already explained the average annual precipitation over most of Sinai is less than 100 mm and in many areas is even less than 50 mm. Thus, even a humid spell which might have doubled or even tripled, this quantity would still have produced only fairly low precipitation. Moreover, a dry spell may have occurred after the tribes penetrated the desert. The tribes may thus have arrived in the area expecting to find flowing springs, to be surprised by dry rock. In such cases, the most advisable thing would have been to dig into the rock to locate the fractures which supplied the water to the springs. These fractures might have contained water which would then fill a shaft excavated into the rock (Issar and Gilad 1982) (Fig. 15.1).

This procedure, which most probably was also known to the ancient inhabitants of this area, is now applied by the local Bedouin tribes occupying the crystalline province of Sinai. The procedure, which I learnt from them is the following:

1. Locate a dried out spring by looking for a white crust (namely travertine) filling the fractures in the hard (crystalline) rocks.
2. Once such a crust was found on a fracture look whether there iare bushes or trees along this fracture.
3. Once you find such a fracture follow it to the point where it is traversed by a soft decomposing black rock or clay (In geological terms where the hard granite or porphyry rock is traversed by soft gabbroid or diabase dike and, its upper part already decomposed into clay).
4. Dig a shaft into the decomposed rock while take care that the shaft follows the fracture.
5. At a certain depth the water will start to seep from the main fracture, and small fracture in the walls, into the shaft. In some places it is necessary to remove the lime deposited in the fracture to enhance the flow.
6. Ask your colleagues to draw the water, by a bucket and rope to enable you to dip the shaft until the water reaches your waist.
7. Bless Allah for the water he gave, and start to build a trough to water your goats.

Fig. 15.1 An orchard in the mountains of southern Sinai, irrigated by the water flowing from the fractures of the rocks. (Photo by A. S. Issar)

Thus, the story of the striking of the rock to obtain water is not altogether alien to the situation in southern Sinai. It would undoubtedly have looked miraculous to people coming from a land where the only source of water was the big river, the canals dug along its banks and shallow shafts dug near by.

The question of the location of Mount Sinai, discussed in the preceding chapter involves the question of the place where Moses smote the rock; while in the book of Exodus (17:1, 17:6) it is at Rephidim, in the book of Numbers (20:1, 20:11) it occurs at Kadesh. From the story of the wanderings it looks as if it took place on the route from the place of crossing from the western to the eastern edge of the Sea of Reeds, namely through the central part of Sinai.

Still it is possible that a few of the tribes reached the southern part of Sinai and confronted the problem of striking the crystalline rocks building this region, in order to get water. If the events of the striking of the rock took place in central Sinai it was most probably a limestone rock which builds the anticlines of northern Sinai.

As mentioned, the site where Miriam the sister of Moses passed away, and where the incident of striking the rock instead of just speaking to it, were at Kadesh. From here the mission headed by Joshua and Caleb went to spy out the Promised Land, and from here the first attempt to storm the land was made (Numbers 13:1–26). Kadesh Barnea is probably the present site of Ein Kudeirat (Fig. 10.1) in eastern Sinai; indeed, there are not many sites which fit the biblical stories about this place. In the first place, it is an ideal concentration place for a few tribes to camp in the desert. A large spring issuing from the limestones rocks of Avdat Formation of Eocene Age, emerges at the head of a valley. This spring is fed by the rains and floods flowing over the rocks building the anticline of the Ramon, which is the highest mountain of the Negev (reaches 1000 m above MSL) and thus receives more precipitation than the lower surroundings. The water infiltrates into solution channels in the limestone rocks until it reaches impermeable chalk layers of Lower Eocene age and marls of Paleocene age, on which a regional perched water table is formed. In addition to the main spring of Kadesh Barnea the special geological conditions cause many small springs to seep from the rocks in this area. One of them still bears the name Ein Qadis.

The outlet of the narrow valley of Kadesh Barnea is into a broad valley where the small town of Al-Quseime lies. Here, many small brackish springs seep from the ground allowing many shrubs, palms and grasses to grow. Today the spring of Ein Qudeirat flows at about 40 cubic meters per hour. This quantity may suffice to supply drinking water for a few tens of thousands of people providing they water their stock from the brackish springs in the lower valley. Taking into account that a more humid period would have caused an even larger flow, one can understand the decision to make Kadesh Barnea the pivot point for the wandering tribes.

It is more difficult to understand the story about the need to strike the rock to get water at Kadesh. Was the story from another place transferred to the more permanent location of the tribes or did the spring of Kadesh also fail, thus necessitating excavation into the rock? According to records known to date, this spring has never dried up even after a series of a several dry years, though its flow did diminish. The possible explanations thus are either a more conservative one, namely that the last stages of the stay at Kadesh Barnea, a dry period restarted, which caused the springs to dry up and which necessitated the digging of wells. The same dry period forced the Israelites to try and break out from the desert into the more humid lands. The less conservative approach will suggest that this story was attached to this place by the writers and editors of the Bible in some later period. As mentioned several times already, the purpose of the present study is not to explain Biblical and mythological phenomena but to offer a better understanding of the environmental background which may have caused them to happen and be remembered in one way or another. Thus, within the boundaries of this assignment, it can be claimed that, while with the better knowledge of the Bedouins' methods of excavating for water in the crystalline rocks, the story of Moses striking of the rock can be understood, it does not fit so nicely with the hydro-geological situation at Kadesh Barnea.

The base at Kadesh Barnea saw the first attempt of the Israelites to enter Canaan by force and their first defeat. They were beaten back by the King of Arad (Fig. 15.2). This story raises some questions. The archaeological excavations at the site which was definitely identified as the ancient town of Arad show that this town was flourishing as a strong fortified city during the Lower part of the Early Bronze Age. It was abandoned at about 2650 B.C and was not rebuilt until ca. 1150 B.C. Thus during the Late Bronze Age, from ca. 1250 to 1150 B.C., the assumed period when the Israelites arrived, the town was abandoned (Aharoni and Amiran 1964; Issar 1976).

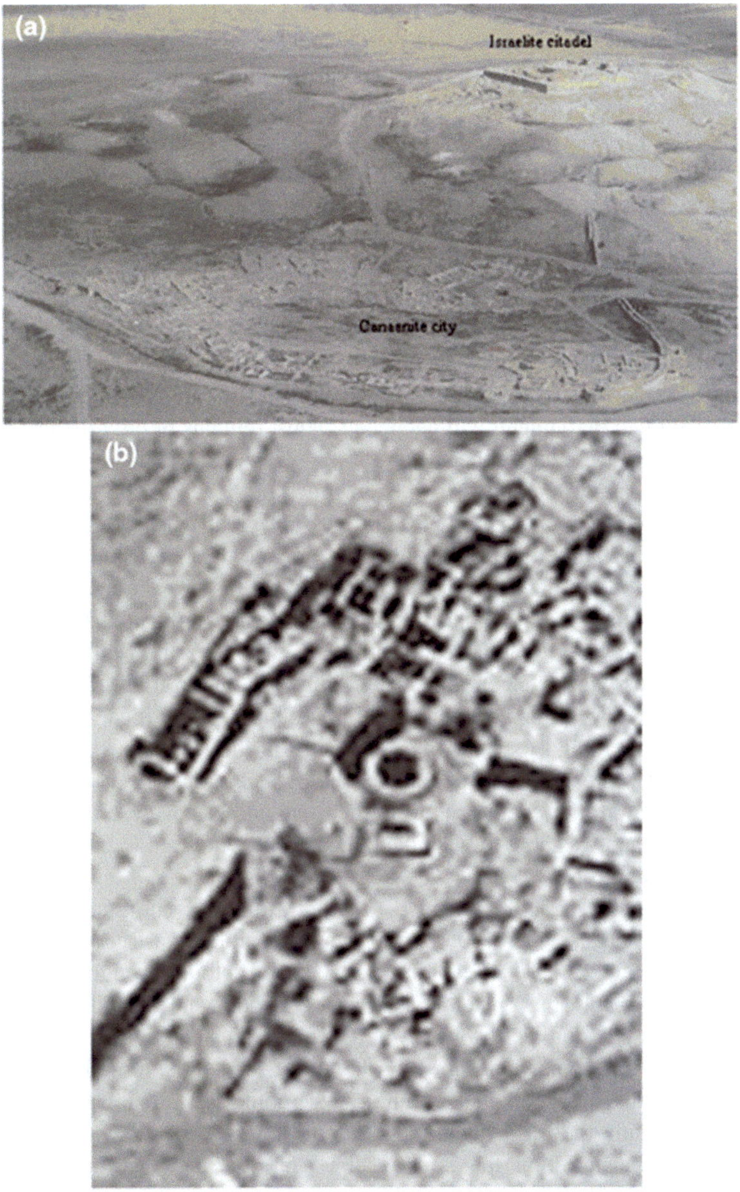

Fig. 15.2 **a** Air photo of Tel Arad **b** Air photo of central well—Canaanite City—Tel Arad

In the opinion of the author of the present work, there exists no real controversy between the story of the Bible and the archaeological findings. In general terms, it can be said that the history of this site as revealed by the excavations fits well with the paleo-climatic story as seen in Figs. 1.2 and 1.5. The flourishing of the city during the time of the Early Bronze Age coincides with the humid climate characteristic to this period, while its decay and abandonment at about 2600 B.C. is in agreement with the a period of desiccation of the climate, which can be seen even on these curves.

While the city of Arad was still settled, its people excavated a deep shaft in the lower stretches of the city reaching into a shallow groundwater table. Although the archaeologist Ruth Amiran, who excavated the site, maintains that the well is from the Iron Age (personal communication), the present author disagrees with this conclusion, as he maintains that the archaeological considerations did not take into account the fact that a humid spell could have caused the water table to rise and thus in the early stage a rather shallow well could have sufficed for the town's water supply. During the Iron Age when methods of digging wells improved, the well was enlarged in diameter and deepened. Thus, remains of only the last period, namely Iron Age, were left at the bottom of the well.

This well ensured a supply of water in times of siege. As the area replenishing this water table is rather limited and does not greatly exceed the area of the town, it seems probable that a series of dry years and the dense building up of the replenishment zone brought about the drying up of this project and the desertion of the city by most of its people. Spells of wet years during the Upper Bronze period caused the water table to reappear and, as the site of the well was known to the semi-nomads who dwelt in the surrounding area due to tradition passed from father to son, they would have again gathered around it. Thus, it can be understood that the site of Arad and its well, like other sites in the vicinity where other shallow wells exist, were settled by people headed by a chief, called the King of Arad. This chieftain controlled the passes from the Sinai and Negev deserts to the mountain of Hebron and was able to push back the people who tried to pass through the land under his control.

At this stage, the Israelites decided to avoid a direct assault across the desert border with Canaan, where too strong an opposition was encountered, and to circumnavigate it. They tried to make a short detour through the Land of Edom, most probably the present Eastern Negev, but were denied free passage. They then pressed southward toward the Gulf of Eilat namely the Sea of Reeds (Yam Suf) crossing somewhere north of the gulf, to reach the more humid heights of the mountains of Moab and Ammon, where they avoided war by abstaining from harassing the local inhabitants whom they considered kin. They moved northward, defeated Sihon the king of the Amorites and Og the king of the Bashan, enabling the Israelites to inhabit the Gilead and Bashan heights and to be ready to cross the Jordan. It is interesting to note that after they had crossed the rift valley and gone up the mountains of Trans-Jordan, no more complaints of thirst were reported. During this part of their journey the tribes experienced an event connected with the excavation of a well for which occasion a special hymn was composed.

> Spring O Well, sing ye unto it. The princes digged the well, the nobles of the people digged it, by the direction of the lawgiver, with their stave. (Numbers 21:17, 18)

References

Aharoni Y, Amiran R (1964) 'Arad' Archaeol 17:43–53
Issar A (1976) The Evolution of the ancient water supply system in the region of Jerusalem. Isr Expl J 26(2–3):130–136
Issar A, Gilad D (1982) Groundwater flow systems in the arid crystalline province of southern. Sinai Hydrol Sci 27:309–325

MIX
Papier aus verantwortungsvollen Quellen
Paper from responsible sources
FSC® C105338

If you have any concerns about our products,
you can contact us on
ProductSafety@springernature.com

In case Publisher is established outside the EU,
the EU authorized representative is:
Springer Nature Customer Service Center GmbH
Europaplatz 3, 69115 Heidelberg, Germany

Printed by Libri Plureos GmbH
in Hamburg, Germany